A Field Guide to

COMMUNICATION

A Field Guide to

COMMUNICATION

For Students in the Humanities and Social Sciences

Pamela Shaw

OXFORD
UNIVERSITY PRESS

OXFORD
UNIVERSITY PRESS

70 Wynford Drive, Don Mills, Ontario M3C 1J9
www.oupcanada.com

Oxford University Press is a department of the University of Oxford.
It furthers the University's objective of excellence in research, scholarship,
and education by publishing worldwide in

Oxford   New York

Auckland   Cape Town   Dar es Salaam   Hong Kong   Karachi
Kuala Lumpur   Madrid   Melbourne   Mexico City   Nairobi
New Delhi   Shanghai   Taipei   Toronto

With offices in

Argentina   Austria   Brazil   Chile   Czech Republic   France   Greece
Guatemala   Hungary   Italy   Japan   Poland   Portugal   Singapore
South Korea   Switzerland   Thailand   Turkey   Ukraine   Vietnam

Oxford is a trade mark of Oxford University Press
in the UK and in certain other countries

Published in Canada by Oxford University Press

Library and Archives Canada Cataloguing in Publication

Shaw, Pam
A field guide to communication: for students in the humanities
and social sciences / Pam Shaw.

Includes bibliographical references and index.
ISBN 978-0-19-542513-0

1. Communication in the humanities—Textbooks. 2. Communication
in the social sciences—Textbooks. I. Title.

H61.8.S43 2008   808'.0663   C2008-905862-3

Cover image: Urban Cow/iStockphoto

This book is printed on permanent acid-free paper ∞.

Printed and bound in Canada.

1 2 3 4 — 12 11 10 09

Contents

List of Figures

List of Boxes

Preface

A Field Guide to Communication is a practical, comprehensive source of information on developing skills in writing, research, presenting, graphics, and mapping for university students in the humanities/social sciences. The text is presented in four sections: 'Writing Skills', 'Research Skills', 'Presentation Skills', and 'Illustrations and Mapping'.

Part I (Chapters 1–5) instructs students in writing with clarity and precision in both academic and professional environments. While both environments require writers to have a clear understanding of the topic at hand and to ensure their writing is grammatically correct, there are key differences between these two styles of writing. However, the comparison between the two is not to suggest that one is superior to the other, only that the senior student and the professional should be able to switch easily between the two styles, depending on the requirements of the assignment.

Within each style of writing, the formats likely to be encountered by the student and new professional are reviewed. For the academic, formats including abstracts, research proposals, and dissertations are reviewed. For the professional, press releases, newsletters, and briefing notes are among the formats considered. The objective of the chapters in Part I is to illustrate that a good writer will be able to produce clearly written, informative information in any required format. Finally, Chapter 5 explores ethics in writing, how and why we need to give credit to the sources of ideas not our own, and the issue of plagiarism.

Part II (Chapters 6–8) focuses on the basics of research—tailoring research approaches to the project, the use and misuse of sources, practical information on citing sources, and, in Chapter 8, writing field reports, which aims to assist the student in taking a more comprehensive and holistic approach to understanding and recording field observations.

In Part III (Chapters 9–13) we look at verbal and non-verbal presentation skills. This part begins with a technical review of the elements of public speaking, including pitch, pace, tone, inflection, and pause, then shifts to the practical aspects of developing improved presentations. Aspects of presentation discussed here include:

- reading the audience, with information on studying facial expressions and body language;
- shaping audience behaviour;
- tailoring the presentation to the audience;
- dealing with difficult questions and audience members;
- configuring a room to your presentation;
- developing a personal presentation style.

Part IV (Chapters 14–17) considers the importance of 'rules' in graphics, as well as the basics in document layout and design. These graphic rules—and when they can be broken—are then applied to figures, tables, static presentations (such as posters at a conference or display boards at an open house or public meeting), and PowerPoint presentations. Information is also presented on the rapidly evolving presentation formats available through new technologies, such as streamed videos and RSS feeds.

In Chapter 17, the last chapter in this part, we examine the value and use of maps. The power of the map to communicate data is considered, as are the basic elements of mapping. While this chapter will address the use of geographic information systems (GIS) in mapping, practical instruction in GIS is left to other authors. Computer mapping is a specialized field undergoing rapid change. This book examines the product of GIS—the map, the graphics, the presentations—and considers the use of this technology in representing 'truth' in mapping.

A Field Guide to Communication is intended to provide students with the information needed for professional positions in the humanities/social science fields in an inclusive single volume. It should be an essential guide for students who want to improve their skills in writing, research, presentations, graphics, and mapping.

Features

- Provides a comprehensive introduction to communication for the humanities/social sciences student—graphics, professional and academic writing, research, and presentation skills.
- Presents practical, tested, and focused advice on improving communication abilities.
- Provides specific tips for improving the speaking voice.
- Increases the student's understanding of how information is controlled and manipulated through effective visual presentations.
- Website access is available to students and instructors. The instructor's section includes PowerPoint presentations for each chapter, with all graphics in a downloadable format, an exam database, an instructor's workbook with suggested assignments and exercises for each chapter, and links to relevant websites. The PowerPoint presentations, instructor's workbook, and exam database are also available to instructors on a CD-ROM. The student section on the website contains practice quizzes, links to relevant websites, and practical examples of assignments.

Acknowledgements

I am grateful to Jim, Lindsay, and Lauren for their patience through this long endeavour; to Dr. Don Stone, Professor Emeritus, Vancouver Island University, for his encouragement; and to Dr. Larry McCann, University of Victoria, for setting standards of excellence I someday hope to achieve. I would also like to say thank you to Peter Chambers of Oxford University Press for his sage advice: the completion of this text is a tribute to his tenacity. My heartfelt thanks to you all.

PART I

Writing Skills

Part I provides information on writing with clarity and precision in both academic and professional environments. While both environments require the writer to have a clear understanding of the topic at hand and to ensure his or her writing is grammatically correct, the expected formats vary greatly. Here, we examine the formats often encountered by students at universities, and learn something about writing abstracts, term papers, literature reviews, and proposals. The importance of writing in an academic style is considered, and the student is provided practical tips on improving the writing depth.

Also examined in this first part are the formats encountered in a professional context, including proposals, press releases, newsletters, briefing notes, resumes, and cover letters, along with recommendations on how to achieve precision in professional writing. The final chapter in Part I looks at the various issues surrounding ethics in writing, including sources, citations, and plagiarism.

Chapter One

Write to Communicate

Since humans first began to use symbols to represent objects, and later grouped these symbols to convey ideas and meanings, the written word has shaped the distribution of knowledge. The ability to disseminate ideas without physical contact revolutionized learning; from the first book that rolled from a printing press, intellectual thought has been shaped by the ability to communicate through writing.

The primary intent of professional writing is to communicate with the reader. While other forms of writing may be intentionally complex, professional writing seeks to persuade or educate the reader with language and grammar that ensure the reader understands the meaning of the text. A successful professional writer is one who gets the message across—there is little point in producing text that is incomprehensible to the target audience.

This chapter and the following chapters in this section consider techniques to ensure the 'readability' of the writer's work. As well, practical information on presentation formats that will be required of the professional writer are presented here. The ability to write clearly is a critical skill—with some effort and applied knowledge, the graduating student can make the transition to employed professional with style.

Plain Language: What Does That Mean?

Plain language is writing that considers the needs of the reader first. The writer makes every effort to ensure that the reader will be able to fully understand the information as presented through clear sentence structures, restricted use of jargon, and restricted use of unnecessarily complicated vocabulary. Graphics or other pictorial information may be used to better explain information, and material is presented at the level of the target audience.

There are several benefits to writing in plain language:

- The reader can understand the information presented.
- One reading (instead of many) is required to digest the information.
- The reader's reaction can be better predicted by the writer.

- The chance of error is lessened as the writer must pay careful attention to the content of the piece.
- There is less of an opportunity to be misunderstood.

Plain language is not a 'dumbing down' of information; instead, it takes a clever writer to present a message in simple, concise language that speaks to—not down to—the reader. Plain language uses all the resources available to the writer so that the message is wholly understood by the reader—grammar, word choices, sentence structure, and even the typeface and graphic style used all are aspects of writing in plain language.

The call for writing to be understood has been an active area of criticism for the last century. In response to what he saw as the unnecessary complication of the written word, William Strunk Jr published *The Elements of Style* in 1919. Strunk called on writers to 'omit needless words' and 'use definite, specific, concrete language' with a focus on communication, not complication.[1] The text, also referred to as 'Strunk and White' or 'the little book', has been revised slightly through four editions (the most recent in 1999) but remains an essential book for academic or professional writers seeking to improve the clarity of their work.[2]

Others have continued Strunk's quest for the production of clear, precise, unambiguous writing. George Orwell added to this discussion in his famous 1946 essay, 'Politics and the English Language'. In it, he criticized government for complicating communications unnecessarily, stating: 'In our time, political speech and writing are largely the defense of the indefensible. . . . [P]olitical language has to consist largely of euphemism, question-begging and sheer cloudy vagueness.'[3] Orwell railed against the distortion of meaning created by overly written text and questioned the motivations of individuals writing to deceive. His own writing remains a model of a clear, succinct writing style, following Strunk's rules for clarity and focus.

The shift to writing in plain language has been gathering momentum since the early 1980s in North America, and professional organizations have been founded to assist in disseminating information on writing in plain language. Canadian provinces have rewritten legislation to improve readability (for example, compare Alberta's current Local Government Act to its predecessor, the Municipal Act, and see Box 1.1 for changes in the use of words as recommended by Nova Scotia's provincial government) and many agencies employ professional writers to ensure that published information is written in an easily understood manner. In Europe, several members of the European Union have begun a 'Fight the Fog' campaign to require bureaucrats to publish only clearly written documentation in publications produced by this new government organization. The Directorate-General for Translation at the European Union has published a guidebook for writers and translators, describing how to best present works in English in a plain language format; they provide suggestions intended to assist the writer in 'mak(ing) sure your message ends up in your readers' brains, not in their bins.'[4] This brief guidebook presents three methods for 'fighting the fog'[5] and writing in plain language.

Box 1.1 Nova Scotia's Plain Language Initiative

The province of Nova Scotia has published a list of words to be avoided when writing in a plain language style. They suggest the following replacements:

Instead of . . .	Use . . .
acquire	get
aforementioned	previously mentioned
allocate	give
at the present time	now
comprises	is made up of
denote	show
for the duration of	during
empower	allow
endeavour	try
expeditiously	as soon as possible
facilitate	help
henceforth	from now on
nevertheless	even so
obtain	get
procure	get
regarding	about, on
render	make, give
undertake	agree, promise
utilize	use

Source: For a complete list and more information on Nova Scotia's plain language initiative, see <www.gov.ns.ca/cmns/plainlanguage/words.asp>.

1. *Don't state the obvious.* This raises an important issue—the need to know your audience and what would be obvious to them. There is a balance between 'don't state the obvious' and providing enough information for the reader, and this balance point is found when the writer understands the needs of the intended audience. If you are an anthropologist writing for a peer-reviewed academic journal, it is likely that the intended readers will be knowledgeable and need little explanation on well-known concepts and principles. However, if you are writing on the same topic for a local newspaper, the level of

knowledge among readers will vary widely. Attention must be given to finding that point between too much explanation and too little information.

2. *Don't clutter your text with redundant expressions like 'as is well known', 'it is generally accepted that', 'in my personal opinion', 'and so on and so forth', 'both from the point of view of A and from the point of view of B'.* Once a writer understands plain language, this point can be addressed with little difficulty. Look for language traps in your own writing: is there a particular word or phrase that you overuse? Does every sentence contain a filler phrase like 'as stated previously' or 'it would seem that'? As you read over your work, do you see a clearer way to make a point? Again, the removal from clutter in writing is an issue of balance. Do not remove so much content that your product becomes a bulleted list, but be cognizant of words that are nothing more than filler. Consider what the reader needs to concentrate on, and plain language will flow.

3. *Don't waste words telling readers what the text is going to say, or reminding them what it said earlier. Just say it. Once.* This can be difficult to achieve, as some writing formats require the writer to restate key points. For example, if a research report calls for an abstract, introduction, analysis, and conclusion, there will be overlap in language and ideas among these sections. Similarly, in lengthy pieces of writing, such as theses, longer journal articles, and books, the writer will, on occasion, need to remind the reader of what has gone before and how it relates to the present specific topic or idea. What needs to be avoided, however, is the unnecessary rephrasing of the same point, presented over and over. Again, this is an issue of balance: write in the required format, but ensure that each sentence says something interesting and new.

While government agencies and academics are most often faulted for not using plain language, many private industry organizations also produce writing so thick as to be incomprehensible. In recognition of the need to ensure that readers understand what they are signing for, buying, or committing to, many organizations are adopting plain language practices for their contracts and customer agreements. For example, the Canadian Bankers Association has produced a model document to be used by banks to ensure that bank clients understand what they are signing (see Box 1.2).

When writing in plain language, the writer should limit the use of jargon or acronyms that cannot be easily understood by the average reader. While most readers would recognize standard abbreviations such as 'm' for metre or 'ha' for hectare, most abbreviations are specific only to a group of individuals with shared interests or collective technical knowledge. That is, most members of the CIP and particularly the PIBC would understand that the BCLGA sets out requirements for OCPs, DPS, DVPs, and the BOV; other readers would need more information to follow a document filled with these acronyms.[6]

In addition, the same acronyms may be used by different organizations with very different meaning—consider the meaning of 'PGA' to members of the Potato Growers' Association versus the Professional Golfers' Association. Indeed, in one famous instance, the World Wrestling Federation (WWF) was forced through

---

Box 1.2 Canadian Bankers Association: Plain Language in a Mortgage Contract

Before	After
It is agreed that in case default shall be made in payment of any sum to become due for interest at any time appointed for payment thereof as aforesaid, compound interest shall be payable and the sum in arrears for interest from time to time, as well after a before maturity, shall bear interest at the Charge Rate, and in case the interest and compound interest are not paid on the next interest payment date after the date of default a rest shall be made, and compound interest at the rate aforesaid shall be payable on the aggregate amount then due, as well after as before maturity, and so on from time to time, and all such interest and compound interest shall be a charge upon the Charged Premises.	If a regular mortgage payment is late, we calculate the extra interest you owe for being late every day, using the annual interest rate of your mortgage. You pay interest on both the principal and the interest portion of the payment that is late. When we receive a payment, we will deduct the interest charges for the late payment and the interest owing on the principal amount first, before any part of the payment is applied to reducing the principal amount. We may also decide to apply the late payment to other amounts you may owe, for example, property taxes.

Source: Canadian Bankers Association, at: <www.cba.ca/en/viewdocument.asp?fl=3&sl=11& tl=127&docid=296&pg=1>.

---

legal action to change its name to World Wrestling Entertainment (WWE) because the acronym it shared with the much older World Wildlife Fund (WWF), a non-profit environmental organization, caused confusion among potential donors to the latter group. When using acronyms, be certain that the reader understands your meaning and consider any possible references that may be confusing or distracting. Also, if an acronym is used, it should follow the first use of the term in the document—that is, the first time 'Professional Golfers' Association' is written, the acronym '(PGA)' should follow in parentheses. The acronym may need to be defined more than once in a document, particularly if the document is in sections or chapters, and the reader may skip past the section providing the definition; alternatively, a list of abbreviations may be included following the contents page to obviate the need for repetition.

Depending on the topic, jargon or technical terms may be required to fully explain an idea. For example, an article written for geologists may reference terms like 'highly foliated metamorphic' or 'decomposed igneous' when discussing types of rocks, and might follow with a discussion comparing the merits of Rossi-Forel versus modified Mercalli scales for the measurement of earthquakes. These terms would be understood by the geologists, but are largely meaningless to the general public. Depending on the audience, definition of the terms may be required, or the writer might consider if they are needed at all—there may be a better, clearer way to express ideas and concepts. Often, the addition of descriptive text or graphics helps the reader to understand more fully the writer's intentions. The idea is to write to be understood by the target readers with a view to maximizing their understanding of the written material.

We write to inform, explain, persuade, or to cause the reader to think about a particular issue or topic. Good writing communicates a message from the writer to the reader with clarity, style, and in a way that cannot be misunderstood. The use of plain language is a foundational part of ensuring that the reader understands your message.

Chapter Review

This chapter reviews a movement in writing called 'plain language' where a document is written to meet the needs of the reader. Plain language is not a 'dumbing down' of information, but a way of writing that ensures the reader will understand the information presented. As a writer, consider if the word choices, grammar, and sentence structure are appropriate for the target audience. The format of graphics and illustrations and even the font used can add to the readability of the document. Use plain language as a means of making your writing more accessible and to ensure that your message will be understood.

Review Questions and Activities

1. List five benefits to writing in plain language.

2. In *The Elements of Style*, Strunk called for writers to _____
    _____.

3. Review *The Elements of Style* and select five ideas for improving writing. Using a recently completed term paper, review and edit the paper to implement the five selected improvements.

4. Using on-line e-journals, seek out a section of academic writing that is not written in plain language (select a section of text of 200–300 words). Note the techniques, word choices, or structures that make the writing unnecessarily complex, then rewrite the section in plain language.

5. Using on-line sources, seek out a section of business or government writing that is turgid rather than plain, and rewrite it in plain English.

Chapter Two

Academic and Professional Writing

This chapter provides a number of methods to achieve clarity over ambiguity and concision over wordiness in academic and professional writing. The division into two writing styles—academic and professional—is in no way meant to suggest a dichotomy between these styles. Professional writing and academic writing are two different formats that you, as a student and professional, are capable of writing in and are prepared to use, depending on the circumstance or the assignment.

As a student, much of your writing to date has likely been essays and term papers. The parameters of your paper are often established by the professor (10 pages, double-spaced, 5-centimetre margins) or by the university (with detailed formats that must be adhered to for theses and dissertations to fit national standards). But beyond the technical format of academic papers, academic writing also calls for a certain style, a way of writing that is specific only to the academy. Academic writing calls for the writer to show that she has a level of knowledge and

---

Box 2.1 Is Kansas Flatter Than a Pancake?

In a 2003 article in the *Annals of Improbable Research*, three researchers used digital image processing and a confocal laser microscope to collect elevation points on a pancake and on a piece of Kansas (a site similar in size to a pancake, and located near Wichita). After some analysis, the researchers proved that, indeed, Kansas is flatter than a pancake. Their writing assumed an academic style for research that could be considered perhaps frivolous in nature. However, this article shows the power of academic writing to add legitimacy and depth to a research question. By writing in an academic style and in a research journal format, the writers created a work both memorable and informative.

Source: M. Fonstad, W. Pugatch, and B. Voght, 'Kansas is flatter than a pancake', *Annals of Improbable Research* 9, 3 (2003): 16–18.

multi-layered understanding of a selected topic. The audience for academic writing is your professors, fellow students, learned professionals, and academics in other institutions—in short, academic writing is read by a highly informed audience with knowledge of the topic and high expectations for accuracy and depth.

Generally, academic writing is serious and intended to prove a hypothesis or justify an argument. In some cases, the writing is presented so that other researchers can verify results and potentially reconstruct the experiments discussed. Academic writing calls for the writer to be accurate and thorough in the discussion and analysis of complex topics. Good academic writing is not wordy and ambiguous; it is comprehensive, detailed, and clear, but requires a reader who is prepared to accept the challenge of understanding the depth and breadth of discussion developed through the writing.

In professional writing, the objective is generally to persuade or inform the reader. It is not necessary for the writer to prove a thesis or justify a research program; instead, the writer seeks to convince the reader of a viewpoint or inform the reader on a particular topic. Professional writing ranges from advertising copy to staff reports: the purpose of this writing format is to attract the readers' interest and draw them towards an intended conclusion.

Both styles of writing are reviewed in this chapter, with a focus on some common formats likely to be encountered by the student and professional. First, however, some common approaches to 'good' writing are outlined.

Approaches to Academic and Professional Writing

Pre-planning

We write so that someone will read our work and understand what we have to say. Both academic and professional writing will benefit from absolute clarity on the purpose of a particular piece of writing and the goal(s) to be achieved.

An important first step, and one that cannot be neglected, is to determine the *purpose* of your document. Ask yourself:

- Why are you writing this?
- Who will be reading it?
- What is the point of the written communication?

If you have the luxury of determining your own topic, find one that is current, relevant, and that you find interesting. However, it is more likely in both academic and professional writing situations that you are assigned a topic with little consideration as to your level of interest. If you are working for a local government, a manager may ask you to write a staff report on fish farming one day and heritage conservation the next. In an academic environment, depending on the course, your professor may require 10 double-spaced pages on issues associated with an aging population or types of sedimentary rocks. In any of these situations, your first task is to be clear on the assignment and the purpose of the work. Ask yourself the above

Box 2.2 The Successful Student Becomes the Successful Professional

Do you ever think about what are you doing in university?

You are learning:

- to be responsible for your own learning—to figure out what is true and relevant;
- how to adapt to different instructional styles;
- how to conceptualize a problem or issue—going beyond the obvious;
- better reading skills;
- better writing skills;
- how to work to a format—theses and dissertations (and often term papers) have very specific technical requirements;
- how to find information;
- how to find your own viewpoint and form an argument that builds on a foundation of relevant research.

You are learning rigour in style and technique, and in using the work of others (your references), to build an argument, to draw connections among ideas, and to understand the reasons behind specific conclusions or outcomes.

In short, a student who succeeds at university will transfer learned skills to a professional work environment and will use these skills to succeed (an important point to keep in mind as you complete yet *another* research paper).

three questions each time you have a new assignment. Be clear on the purpose of the work before launching into research or writing—you will save valuable time if all your efforts are directed towards a clearly defined purpose.

In this pre-planning phase, it is also important to take some time to brainstorm before locking on to a particular point of view or conclusion. While your initial idea may be valid, allow yourself the opportunity to expand your thinking beyond the usual or obvious. This is a critical component that is often skipped over in response to tight deadlines; however, even if you take only an hour or two to consider new possibilities, the effort expended may reveal a whole new direction that would otherwise have remained uninvestigated.[1]

Practical tip. Right from the pre-planning stage, document your sources. It is frustrating to work backward to try to find the author of a particular quotation or a reference for key information. The few seconds needed to accurately record a reference can save hours of wearisome searching. This documentation need not be cumbersome: many universities have adopted on-line referencing systems—see the section on 'Using Sources and Citations' later in this chapter.

Outlines

Do an outline! As painful as the development of outlines has been through your academic career, they are an invaluable tool for moving forward as a professional. A well-developed outline keeps your writing on track and saves time in researching as the process becomes one of filling in the outline instead of creating structure. In a professional environment where you are required to produce many reports in a specified format, a good outline can develop into a template that allows you to fill in factual material quickly and then focus on the analysis or recommendations as required.

An outline lets the writer logically follow an idea to completion—if the outline does not make sense, neither will the final document, and it is easier to adjust a point-form outline than to realize you have spent many hours researching material that has no relevance to your final product.

Look for flow in the outline—build each section on the previous one to reach a logical and complete conclusion. The outline should show a rational progression, tracing a line of thought across time or topics.

The outline may be in point form or written in complete sentences. Each format has its advantages: the point-form outline is simple and will develop more quickly, while an outline structured in complete sentences or paragraphs is more easily developed into the final document. Each requires a different approach—a good sentence outline presumes that the writer has a fair command of the topic before beginning the research, while a point-form outline works when a paper is being developed from a lesser knowledge base. Overall, the objective is to create a complete document—with a beginning, middle, and end—that flows in a logical format and meets a specified purpose.

There are two ways to do an outline: *linear* and *circular*. A third method *combines* aspects of linear and circular outlines.

A **linear outline** is a means of organizing thinking, ideally on a single page, and should be headed with the title of the paper or report. The *title* is a short statement that speaks directly to the purpose of your paper. Following the title is a *summary or thesis statement*, a short (ideally, one sentence) abstract of what the document is to be about. You need to be able to define your topic in one sentence (or, at most, two sentences). Otherwise, you probably aren't as clear about your topic and what you aim to do with it as you need to be. The outline itself should follow:

1. **Introduction**
   1.1. Expand on your topic.
   1.2. What is your thesis?
      1.2.1. Are you working from a hypothesis?
2. **Background (or Literature Review)**
   2.1. What is known or unknown about your topic?
   2.2. As you do your research, are you finding that the research relates to your topic area?

3. **Headings**
    3.1. Three or four topic areas. These are often common themes in the background research on the topic.
        3.1.1. Group ideas that have a relationship to each other. These are groupings that make logical sense.
        3.1.2. Also use subheadings: these subheadings further break down the main topic areas into component parts.
    3.2. Also keep track of quotations and references that speak precisely to your point or support your arguments.

4. **Conclusion**
    4.1. New information is generally not presented in the conclusion. This section summarizes key points and highlights what the reader should remember from the piece.

In this example of an outline, decimal numbers are used to separate ideas by weight and value (you could also use Roman numerals or the alphabet, or some combination of all three classification systems. You might also use colour coding or symbols to further classify and group ideas). Whatever system is used, the idea is to create a hierarchy and an interrelationship of topics in a logical order. Once you have established your headings and numbering system, you will use this structure to organize your research.

As you organize your outline, consider the weight of each idea. At the major level heading, the largest ideas are represented—and only ideas of equal weight are included at this level. Sub-ideas are represented in descending order, keeping a coherent and logical format. Be prepared to amend your outline as you complete your research—the purpose of the outline is to organize your work, not to force you to discard ideas that do not fit the structure as defined. As you work, refer back to the purpose of your document and consider the target audience. What information would be useful to the audience? What is the most logical way to present this information?

While the outline above considers structuring the document around topic areas, there are other options available for organizing the outline:

- *Chronological.* The document is organized by time progression or by separating a series of events into major stages, such as decades or generations.
- *Compare/contrast.* If the purpose of the document is to understand the similarities or differences among topics, then a compare/contrast strategy can be used to organize the document. One side of the issue is presented, and then the other. Similarities and differences are noted. This format can be useful if a small number of issues are being considered, but can be cumbersome to work through in a multi-issue analysis.
- *Explanatory.* If you are explaining a process or procedure, the logical order is to start with the first step and move towards completion. Ideally, complete information is presented at each step so the reader can follow the logic without referencing back to previous steps. Understanding the needs of the reader is critical in explanatory writing: a writer with extensive knowledge of

a process may miss critical steps that the reader needs information on. Most of us are familiar with directions that provide good initial information, then skip ahead to a conclusion, leaving the frustrated reader searching through partial explanations or overly summarized information. If possible, have a colleague review your explanatory writing to ensure it provides an adequate level of explanation for the target audience.

- *Importance.* The document is organized in a logical order, from most to least important or vice versa. Presenting the most important information first can be useful if you anticipate that you may lose readership as people read through your document (for example, press releases commonly use this inverted pyramid format—see Chapter 4). Conversely, organizing from least to most important can allow the paper to build to an exciting conclusion.

Finally, in creating a linear outline, you might look at the detailed table of contents of your textbooks for examples—these contents pages are, in effect, the final topic outlines that the book authors and their editors arrived at. With a linear outline, you can use topics (words and phrases) or complete sentences, or a combination of the two.

A **circular outline** is a means of diagramming thoughts and ideas and can be just as effective as a linear outline in producing a logical document. Both the idea and connections are represented in the circular outline. In a circular outline, the document radiates out from the central purpose, and colours and connections are used to shape the paper. One format for a circular outline is developed below:

- Begin with a blank page of paper and a half-dozen coloured markers. Ideally, your circular outline will fit on a single piece of paper.
- In black, write the title of your document in the centre of the page, and draw a circle around it.
- In one colour, write the major headings for your document, draw a circle around each one, and link them to the title.
- In another colour, draw lines out from each major heading, and then write each subheading in its own circle.
- Continue the process of branching out until all individual ideas are represented on the paper. Use different colours to give values to the ideas (that is, all ideas at the same subheading level are written in the same colour).
- Consider the use of symbols or graphics (dashed or wavy lines, stars or exclamation points, numbering the circles in logical order) to further match and classify your ideas.
- Consider the connections among the ideas. All ideas should connect, and may cross-link to other circles. If an idea has no relationship to any other circle on the paper, it likely does not belong in the document.
- Continue brainstorming ideas until the outline is complete.

This outline then serves as the structure for your document. Software for circular outlines is also available. Two free programs available on-line are Bubblus (www.bubbl.us) and VUE (vue.tufts.edu).

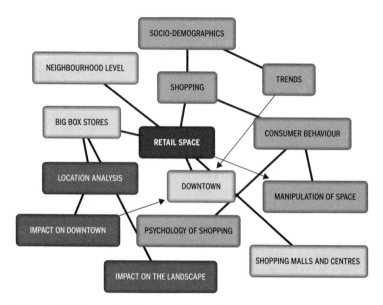

Figure 2.1 Example of a Circular Outline
Source: <www.bubbl.us>.

A third option for outlining is to *combine* the linear and circular styles in a format that works for you. If you are familiar with linear outlines, attempt to use a circular outline for your next written document. Develop your own style—perhaps a linear outline with coloured connection lines or a circular outline with a hierarchy of graduated circle sizes. Find a method that works to organize your thought processes. With practice, an outline will become a valuable tool in producing accurate, effective writing.

An outline is easily prepared in any word-processing program, as sections can be readily cut and pasted to another location. Some writers prefer the 'old school' method of literally cutting and pasting pieces of research together into the outline format, but in the interests of saving a tree, the former method is recommended.

Finally, the cardinal rule for outlines is to maintain consistency in groupings, hierarchies, and order. Create a structure and stick to it.

Drafts

The task now is to fill in the outline. Each piece of relevant research is placed under the appropriate heading, and the document develops from the logical order of your outline. Pick out the salient facts and get rid of the clutter. Pull out any information that blocks the flow of your paper and put it aside. If a piece of information does not fit the structure of your outline, three options are possible:

1. Is this information so important that it warrants an adjustment to the entire structure of the outline? Does it require a rethinking of the purpose of the

paper? If the answer to either of these questions is yes, step back and reconsider the purpose and intended outcome of the document, then make necessary changes.

2. Does the information require the development of a new section? The information may illustrate a gap in the outline. Review the outline to ensure that logical order is maintained.

3. Does the information fall into the category of 'interesting but not relevant'? If information does not fit in the paper, it may be that it does not belong. Edit out the trivia (those interesting but unimportant details that can take a document off on tangents) and keep the document centred on the summary statement.

Revisit the outline often to ensure the document retains shape and focus, and use the outline as a foundation for the document.

It should go without saying, but good grammar is essential for the professional writer. When in doubt, check with a source. Your credibility will be irreparably damaged by poor grammar, sloppy punctuation habits, or weak sentence structures (see Boxes 2.3–2.5). On occasion, you may find that it is necessary to break grammatical rules to best express an idea. However, this rule-breaking should be intentional, not by error or omission. It should be obvious to the reader that the writer knows what he or she is doing and has done so with a purpose.

It should also go without saying, but it is critical to check your spelling. A good writer will adopt a 'zero tolerance' approach to spelling errors. While text messaging and e-mails have created a relaxed attitude in some people towards the importance of spelling in written correspondence (see Box 2.3), this does not hold true in academic or professional writing. At the same time, you should be aware that the spell-checking functions of computer programs are *not* infallible, and misspelled words can be overlooked or, worse, replaced with something that either makes no sense or changes the meaning of your work. A careful review of your work is still required after the spell-check is completed.

As you develop your draft, consider your writing style. Varying the length of paragraphs and sentences will add interest, and the use of short sentences can improve the clarity of your work. Variety in sentence structure will, as well, add rhythm to your work and keep the reader focused. Also, significant information is best presented either at the beginning or end of a paragraph, as the reader tends to place greater emphasis on these sentences. Quotations may be used, but only if they are relevant to the idea being presented (and never to 'pump up' a document with filler).

Bring a sense of completion to your work—there is nothing clever in dancing around an issue that needs to be answered. A weak conclusion can leave the reader wondering if parts are missing from the document.

The draft is complete when the writer finds that there are no new ideas to add, that the information is presented in a logical order, and when a revisit to the summary statement finds that the document responds to the intended purpose of the writing. Unfortunately, many students consider a first draft to be the final version and will submit this as complete. A much more effective approach is to walk away from the draft for a period of time so you can later review the work

Box 2.3 Evolution of Language

Language (both spoken and written) is dynamic and new words are being created at a faster pace than ever before in human history.

- New words are required to label new products and explain new processes (words like 'Facebook'© and 'netspeak' did not exist only a short time ago).
- Words are borrowed from other languages and become part of 'normal' speech and writing (only a short time ago, few would have understood anything about Japanese anime).
- Words shift in meaning (in Shakespeare's day, the word 'weird' meant having supernatural powers. Today, it is more closely associated with being odd or unusual).
- Language evolves, which explains why the works of Chaucer, for example, require concentrated effort to fully understand the meaning and depth of the writing.

Among recent changes is the evolution of text messaging, which is creating rapid changes in written language with new words, abbreviations, and sentence structures—terms like BRB and 4U are widely understood and make sense when used in an e-mail or text message. This does *not* mean that 'netspeak' is grammatically correct or appropriate to use in academic or professional writing. Before using a shortened form of language or using a term that is new or uncommon, consider if the target audience will understand the meaning. Ultimately, we write to be understood, so be certain the meaning will be unambiguous for the reader.

While there has been some linguistic anxiety over the pace of change in language, this evolution has always occurred and is a reflection of human ingenuity in making the language fit our technologies and experiences.

with a fresh approach. If time permits, a 24-hour break is recommended, although deadlines seldom afford this luxury. In a time crunch, even a few minutes spent involved in another task can provide the writer with the necessary mind break to permit a successful rewriting of the document.

Rewriting

In this final step, the draft is improved into a final document. Both content and style are reviewed to ensure that what you have written fully reflects your ability to produce a clear, concise, logical piece that meets the required task.

As you complete the final review, watch for repetition of words that you tend to use over and over. For example, the two sentences that open this section, as originally drafted, included the word 'document' four times, but now the word is used only once. Consider varying sentence structure to break this pattern or

> ## Box 2.4 Good Grammar
>
> The most common grammatical mistakes tend to be those that could be easily captured in a good proofreading of a document—mistakes with homonyms (its/it's, effect/affect, there/their), subject/verb agreement, and sentence structure. Do not skip over proofing your work and never rely solely on spell-check or grammar-check programs—while these programs are useful, they cannot check for meaning and nuance, and certainly cannot detect correctly spelled but incorrectly placed words (for example, *in* instead of *it* or *hover* instead of *mover*.)
>
> A recommended technique for proofreading for grammar, spelling, punctuation, and sentence structure is to read a document backwards (one sentence at a time). The proofreader is not reading for content, but only to ensure that the document is correct; a second proofreading from start to finish would then follow, and would consider the flow and meaning of the document.
>
> This textbook does not contain a section on grammar, sentence structures, and punctuation, as there are many excellent and comprehensive texts available. Keep one on your desk at all times and refer to it often. One recommended text is *The Good Grammar Book* by Michael Swan and Catherine Walter.

replace the repeated word with another term that expresses the same meaning. For example, if you have used 'said' repeatedly, as in 'Jones said that . . .', then 'Smith said that . . .', followed by 'Tom said that . . .', consider replacing the word with any of numerous alternatives—'stated', 'indicated', 'noted', 'suggested', 'exclaimed', 'explained', or vary the sentence structure to break the pattern. The danger in this is the overuse of synonyms—a document can begin to read like a thesaurus if you use a shopping list of synonyms to express your ideas. That is, you do not need to replace every use of 'said' with a different synonym, but if the overuse of a word is impacting the rhythm of your document, then provide some variation in terms or sentence structure to hold the reader's interest.

Other words to watch out for are made-up words that are increasingly popping up in both professional and academic writing. If your document contains terms like 'environmentalize' or 'functionalize', reconsider their use—do they mean anything or add any new nuance to the language? Are they being used to 'intellectualize' the writing (with the unintended opposite effect)?

Also, if you are using a recently created word, are you using it correctly? Every year, dictionaries add new words to the list of terms in common use. In 2006, the *Concise Oxford English Dictionary* (eleventh edition) added the following new words (among many others):

- **agroterrorism** / *n.* terrorist acts intended to disrupt or damage a country's agriculture.
- **mentee** / *n.* a person who is advised, trained, or counselled by a mentor.
- **mzee** / *n.* (in East Africa) an older person; an elder.

- **ponzu** / *n.* (in Japanese cookery) a sauce or dip made with soy sauce and citrus juice.
- **tri-band** / *adj.* (of a mobile phone) having three frequencies, enabling it to be used in different regions (typically Europe and the US).
- **upskill** / *v.* [often as *noun* **upskilling**] teach (an employee) additional skills.

These new words reflect changes in technology ('tri-band'), world influences ('mzee' and 'ponzu'), new issues in society ('agroterrorism'), and language laziness in an attempt to appear trendy and to create new jargon ('upskilling' and 'mentee'). While new words are constantly being created to describe new products and processes, be certain that your contribution to the evolution of English adds to the clarity of language.

In addition, be ruthless if you encounter empty adverbs like 'actually', 'apparently', 'totally', 'wholly', or 'completely' in your writing—more often than not, these terms add nothing to the meaning of the sentence, and in some instances, such as 'very', their use in fact weakens the power of what you intend to say. The word 'totally' has been overused to the extent that it has become a conversational norm, both ubiquitous and meaningless. However, this overuse does not need to be reflected in your writing.

Practical tip. If a word does not add to the meaning of your work, do not use it. The use of 'actually' and 'apparently' have become norms at the beginning of sentences. Ask yourself—do you need these terms? Do they contribute to meaning? Are you using 'apparently' because you are unsure or unclear about the statement that follows? Does 'very' add anything to the power of what you say? Keep your writing razor sharp by avoiding useless filler.

Another tip. Watch for 'weasel words' in your writing, including 'in my opinion' or 'it would seem to be the case that', or again, the overused 'apparently'. Do you believe in what you are writing? Most writing in business requires that the writer know what he or she is talking about. The reader expects that the writer has completed sufficient research or has experience and knowledge to speak with some authority on a subject. While it may be true that a definitive response is not available, a good writer does not rely on 'weasel words' to cover insufficient knowledge or a lack of research effort.

If abbreviations are used, be sure they are initially cited in full and included in a glossary (if the document contains one). If too many uncommon abbreviations are used, it can be confusing for the reader and the reader will quickly lose interest. Also, review to ensure that any citations are correct and presented in an approved format.[2] A final check on your writing is to revisit the thesis or summary statement to ensure that your document successfully responds to this statement of purpose.

Whether in an academic setting or a professional environment, good writing achieves one simple task: the intended reader understands the purpose of the work and the point being made by the writer. The following 10 tips will help to improve the clarity of both academic and professional writing.

10 Tips to Improve the Clarity of Writing

1. *Focus your thinking.* Can you explain your topic in one clearly worded sentence? Do you know exactly what you are doing? If you cannot explain the purpose of your work, it is likely that the reader will have problems understanding the purpose or following the logic of your work. Once a topic has been determined (either by choice or assigned), spend some time focusing in on the topic and be certain you are clear on the purpose of your work. Write out a purpose statement, and test it on a few people who are not familiar with the research area. Even if the topic is unfamiliar to them, does the purpose statement make sense to your test subjects? The development of a clear and concise purpose statement is necessary at the start of any major project (including a thesis or dissertation); referring back to the purpose statement ensures that subsequent research and writing remains on task.

2. *Remove the clutter.* Be ruthless in removing filler statements, extra words, and duplicate sentences. As a student, you are often instructed to produce an essay of a specified length or word count, but your concern should be with content and meaning, not with useless filler. Also, use quotes only if they are relevant, not to pad length. As a professional, you will be expected to make your point as succinctly as possible. Review your work to be certain that the writing is sharp and clean.

3. *Use numbers or graphics.* Could an idea be better represented in numbers or graphics? Sometimes, a number is easier to understand than words; consider the words 'one million dollars' versus '$1,000,000'—the power of many zeros focuses the reader's attention. Numbers imply precision.

   When using numbers, as a rule present no more than three per sentence: a reader's interest is quickly lost to a shopping list of numbers, and long lists of data in a text format are difficult to read and assimilate. Most often, extensive data are better presented in a table or graphic format than in a text format. If you need to present a lot of numerical data on a particular topic, for example, the number and the total square kilometres of provincial and territorial parks across Canada, create a table; in the text, then, you can briefly highlight totals or the highest and lowest numbers when you refer the reader to the accompanying table. (For more on numbers, see Chapter 16.)

   When presenting numbers, carefully consider what is critically important to the reader and if the situation requires numbers (e.g., 51 per cent) or words (e.g., the majority of respondents) to better present the information

Box 2.5 Keep Writing Simple

'Vigorous writing is concise. A sentence should contain no unnecessary words, a paragraph no unnecessary sentences, for the same reason that a drawing should have no unnecessary lines and a machine no unnecessary parts. This requires not that the writer make all his sentences short, or that he avoid all detail and treat his subjects only in outline, but that every word tell.'

Source: William Strunk and E.B. White, *The Elements of Style* (New York: Macmillan, 1979), 16.

to the reader. It should be noted that the meaning of the writing can be manipulated with the selected use of numbers over text: 51 per cent is a more accurate depiction than 'the majority', although both are correct. The selection of number or a word statement depends on the situation and the writer's intentions, but you should never purposefully mislead, which is all too easy to do with numerical data. Consider, for instance, the Canadian federal election of 1988, which was fought largely on the issue of the recently signed Canada–US Free Trade Agreement. One could claim that the Mulroney Conservatives won a resounding affirmation from the electorate for their policy, as the Tories were returned to government with a majority of seats (169 out of 295) in the House of Commons. In fact, however, their victory in this one-issue election was based on only 43 per cent of the popular vote: the Liberals and New Democrats split the anti-free trade vote and together won 126 seats, but their combined percentage of the vote was over 52 per cent.

Generally, numbers from one to nine are spelled out and larger numbers are shown numerically, but this should be considered a guideline only: it is more important to be consistent in presentation throughout a document than rigorously adhering to rules. Again, the most important point is to ensure that the intended audience understands the writing.

Graphics, as noted above, may also be used—a pie chart is a clear way to illustrate a distribution, or a photo may be used to identify a person or place. When using graphics, be sure that the graphic is clearly labelled so the reader understands what is being presented. Graphics and mapping will be considered in greater detail in Part IV of this text.

4. *Keep it simple.* In professional writing, documents are often required to be short, limited, and focused. Academic papers may require a lengthy discussion to fully conceptualize a research question, but this allowance for breadth and depth should not be confused with convoluted writing. Often, the words people remember are those short, powerful statements that speak precisely to the topic—these are the sentences that get noticed.

5. *Do not use unnecessarily complicated vocabulary.* A well-written piece does not require the reader to consult a dictionary or thesaurus to understand it. Use

appropriate words. If a complicated or little-known word must be used to impart the correct meaning, then use it, but if you find you are trying to pump up your writing with unnecessarily difficult text to make yourself appear smarter, you will sacrifice clarity. You must write to share knowledge, not to obscure it, or, as George Orwell put it, 'Write to reveal, not to conceal.'[3]

Read the following paragraphs, written by Joseph A. Ecclesine, and note what is important from a reader's perspective:

> When you come right down to it, there is no law that says you have to use big words when you write or talk.
>
> There are lots of small words, and good ones, that can be made to say all the things you want to say, quite as well as the big ones. It may take a lot more time to find them at first. But it can be well worth it, for all of us know what they mean. Some small words, more than you might think, are rich with just the right feel, the right taste, as if made to help you say a thing the way it should be said.
>
> Small words can be crisp, brief, terse—to the point, like a knife. They have a charm all their own. They dance, twist, turn, sing. Like sparks in the night, they light the way for the eyes of those who read. They are the grace notes of prose. You know what they say the way you know a day is bright and fair—at first sight. And you find, as you read, that you like they way they say it. Small words are gay. And they can catch large thoughts and hold them up for all to see, like rare stones in rings of gold, or joy in the eyes of a child. Some make you feel, as well as see: the cold deep dark of night, the hot salt sting of tears.
>
> Small words move with ease where big words stand still—or, worse, bog down and get in the way of what you want to say. There is not much, in all truth, that small words will not say—and say quite well.[4]

Do you see it? No word in this paragraph is more than one syllable, yet the writer has produced a beautifully written, image-filled text that in no way is lessened by using a simple vocabulary.

6. *Write to your audience.* A good writer can present the same information to a group of university professors or to elementary school students by writing to the level of the audience. For example, if you were writing on the likelihood of a volcanic eruption along the San Andreas fault in the next few years, the content of your paper presented at a conference of geologists would be very different from an article in an elementary school newsletter. However, a good writer should be able to change his or her style to suit any intended audience. A good writer will consider the following:

- Is the vocabulary appropriate to the target audience?
- What would they be interested in?
- How can information be presented to be understood?
- How comprehensive does the article/paper need to be to present the necessary information?

In other words, the good writer writes for the audience, not for him- or herself.

7. *Be honest about bias.* What is your perspective on the issue? It is imperative that the writer be honest about the bias or leaning in his or her writing. Not all writing can be from a neutral perspective (it could be argued nothing ever is), but it is important for you to understand how your perspective has impacted the choice of research materials and the conclusions drawn in the final product (for more on ethics in writing, see Chapter 5; for bias, see Box 2.6).

   Consider the bias of your institution or organization, as well. Sometimes, universities or faculties adopt a particular approach—some universities are renowned for a radical free-thinking approach, while others focus more on the practicalities of graduating employable students. Be cognizant of your university's perspectives on learning, teaching, and administration and consider the influence of this on your own perspectives.

   Within the institution, your role as a student can also be shaped by the biases of others: if you are working on a research project as part of a larger team, it is likely that the research question has been determined by the supervising professor, along with the hypotheses and research methods. Are these perspectives consistent with your own? Are you able to adjust your research to a viewpoint that may be different from that of the rest of the research team?

   As a professional, you may be writing for a publication that considers issues from a predetermined perspective (for example, conservative or liberal, right wing or left wing) or working within an organization that views its actions through a particular lens (that is, perhaps an environmental focus). Before committing to a professional position, research the bias of the organization and determine if it is consistent with your own perspectives. Similarly, when you do research, you need to be aware of the biases of your sources: for instance, the Business Council on National Issues or the Fraser Institute will have a decidedly different view of the development of the Alberta oil sands than will the Pembina Institute or the Sierra Club, and the C.D. Howe Institute will have a different take on public policy issues than will the Canadian Centre for Policy Alternatives.

8. *Test your writing on a focus group.* Ideally, the focus group will be similar to your target audience. Find people who are representative of the readers you are writing for. Ask for a fast reaction to your work. Is the message clear and well presented? Do they have any suggestions on improving the clarity of the text? Be sure that you phrase your questions to indicate that you are concerned with the precision and transparency of your work, not with their ability to understand it.

9. *Be prepared to accept criticism.* There is little point in asking for feedback if you will be unwilling to implement any suggestions. As a professional, you must seek out constructive comments and work constantly to improve your writing.

10. *Analyze the responses to your work.* Often, instead of too much feedback (either positive or negative), we receive little or no response to written materials. However, if you find that people respond incorrectly to instructions in your written correspondence or if they need additional clarification, you may want

to revisit your writing and review it from a plain language perspective. Sometimes a few small changes are all that is needed to increase the clarity of your work. A good writer is never finished improving. See each writing task as an opportunity to produce a better document than your previous work, with greater precision, clarity, and focus.

Box 2.6 Know Your Bias

While the following are overly simplified, the description of each methodology is intended to show how different researchers might approach the same issue. For example, when considering the issue of homelessness, a Marxist might see the problem as a consequence of the capitalistic system. A humanist would attempt to understand the issue from the perspective of the homeless person, while a feminist might consider issues of equity and social justice. Becoming aware of one's own theoretical perspective and the perspectives of other researchers assists in developing a more comprehensive view of any issue or research problem.

Methodology	Perspective
Marxism	The historical evolution of economic affairs and the relationship of groups (classes) to the means of production is central in the Marxist view, generally with a focus on capitalist systems and the role of the individual or collective in power relationships.
Pragmatism	What counts as knowledge is determined by its usefulness. There is no foundational knowledge and knowledge must always be subject to further questioning.
Humanism	Importance is on understanding the place/role of the human as part of the research landscape. Humans are not neutral observers, but part of the events and experiences that take place, and the basic assumption here is the potential for human perfectibility, that is, humans are as good as it gets and we're all in this together, but without a divine will or deity or supernatural Other who is greater than us and cares about (and for) us.
Behaviouralism	The emphasis here is on individual behaviour and the relationship between this and the choices made by the individual.
Structuralism	Humans are part of, and to a greater or lesser extent are controlled by, wider structures such as government, associations, and workplaces, and these structures are both enabling and constraining on human behaviour.

(Continued . . .)

Feminism/ anti-racism	Within feminism, a focus is on gender relationships and well-being, often relating to women and children but also to questions of equity and social justice; anti-racism focuses on many of the same issues, and both feminists and anti-racists seek to eradicate inequality and oppression as it affects those who have been 'othered' by the dominant group or groups within society.
Postmodernism	This relatively recent mode of thought and analysis is suspicious of foundationalism or the search for grand explanatory theories—all truth is partial, pluralistic, relative. The interpretation of an idea or event depends on the observer and his or her place in the social milieu—both self-perceived and as directed by others.
Conservatism	A conservative approach, generally, is more interested in preserving existing structures, traditions, and institutions, is more resistant to change, and is more accepting of limits to personal freedoms.
Liberalism	Conversely, liberalism as a methodology is interested in progress, reform, expanded personal freedoms, and greater individualism.

Chapter Review

This chapter has presented a number of methods to best achieve clarity over ambiguity and concision over wordiness in academic and professional writing. Pre-planning, outlines, drafts, and rewriting are all part of the writing process, and following these approaches in your writing will improve the readability of your work. For both academic and professional writing, the goal for the writer is to be understood by the reader.

Review Questions and Activities

1. Consider your own theoretical perspective. What research methodology do you most closely identify with? Consider the perspectives outlined in Box 2.6. Find out more about three of these perspectives that may represent your personal perspective. Write a short paper (no more than 500 words) that describes your research bias.

2. Select a topic for an academic research paper. Construct both linear and circular outlines. Next, using elements of both models, construct a hybrid that contains elements from each.

3. Select a section from a written work (academic or professional, no more than 100 words). Rewrite the section using no words over one syllable, with no loss in meaning.

4. Research 10 new words that have been included in a recognized dictionary in the last two years. If possible, determine the place/person of origin, and find three on-line examples illustrating that the word is now in common use.

5. Review the last academic paper you wrote for any course. Using the 10 tips for improving the clarity of writing, edit the paper, clearly identifying the places where the paper could be improved.

Chapter Three

Writing Formats for Academics

In your academic career, much of your written work is in the form of essays or major papers. You begin with a question (or set of questions), find references in that field of inquiry, obtain data, complete an analysis, and draw conclusions. Most academic papers follow a standard format established by the professor or the institution. This chapter examines the formats most likely to be encountered by students, and offers practical tips to improve the clarity and focus of academic writing.

Abstracts

More than a purpose statement, an abstract is a summary of the main points of a research paper, academic article, thesis, or dissertation. It accurately abbreviates the contents of the document into a short, concise format that is intended to attract a reader's interest. An abstract typically is no more than 250 words, although individual universities (or professors) may have strict requirements for word counts (as few as 75 or as many as 350) and the elements that must be contained in the abstract (again, be certain of the requirements at your institution).

An abstract is useful in that it forces the writer to synopsize the contents of a much longer document into a series of short sentences that capture the purpose and conclusions of the writing. However, the main purpose of an abstract is to assist a reader in deciding if there is value in reading the entire document, thus, a well-written abstract is important if you want your work to be read by the intended audience. When conducting research, readers may skim hundreds of documents in search of those that are relevant to their research. A well-written abstract provides both a summary of the article and gives some indication of the depth of the article by noting the form of research conducted and the key findings of the document. Ideally, the reader will conclude from reading the abstract that the work is interesting, relevant, and worth reviewing in its entirety.

An abstract typically contains:

- a topic statement (what you are writing about);
- the problem or issue being addressed;

- an outline of the approach to the research;
- research methods used in the document (highlighting original research if it has been conducted);
- key findings.

The first sentence in an abstract states the topic or main purpose of your research. What are you writing about? What is the topic area—does the reader fully understand it from this first sentence? The first sentence may also include the problem statement (what specifically is being researched in this paper), or the topic and problem statement may be two separate sentences. The next sentence will state the approach to the research—your hypothesis or thesis, what the research is intended to illustrate or prove. It may also reference previous work of importance that your research builds upon. This helps the reader (who is familiar with the literature) to place your work in the field of study and quickly gain an understanding of the purpose of your work.

Now that the reader has an understanding of the purpose, problem, and approach, the abstract provides information on the research methods used to study the problem. Again, this is contained in one or two sentences that briefly describe the work—an abstract is meant to attract the reader's attention and summarize the research, not restate every key finding or discussion point.

The final sentence in an abstract restates the main conclusion of the work. What did you discover? Is there a point in the research that you believe is critically important to the reader? The final sentence is the last opportunity to entice the reader into reviewing the entire document.

The abstract is written as one paragraph, single-spaced (even if the requirements for the paper are double-spaced). Formulas and diagrams are not typically included in an abstract.

Often, the abstract follows the format of the writing—the first sentences would be contained in the introduction and background of the paper, the next sentences describe the analysis sections, and the final sentence restates the main conclusion. When building the abstract, consider the flow of the paper and use this to guide the placement of sentences in the abstract.

Ideally, an abstract is among the last components of a major paper completed. While it contains some similarities to an outline (which is completed first), a good abstract explains the written work in such a way that captures the tone and depth of the entire article. While a student may feel that she should complete the abstract first (because it is at the beginning of the document and because it provides a summary of the work), the development of the abstract is better left until the research, analysis, and writing are finished, so that it best represents your completed work.

An abstract balances conciseness in writing with providing sufficient information to attract the reader. Once you have written your abstract, review it first for unnecessary words. Statements like 'at the present time' or 'as illustrated in Chapter Three: Findings' take up space and do not add to meaning: carefully review the abstract to see if it contains filler text that can be removed. Next, consider if each sentence transitions well into the next: an abstract should not be a point-form restatement of the larger work, but a paragraph that tells an interesting story on the key points of the document. Transitional words such as 'In addition', 'Further', or 'Again' may be needed at the start of the middle sentences to pull the reader through the abstract.

Practical tip. If you are quoting your own work in the abstract (lifting a complete sentence out of the document), quotation marks are not required. However, if you are quoting another's work (which would be unusual in an abstract), then normal referencing requirements apply.

To better recognize a well-written abstract, review the abstracts of articles in a journal in your selected field of study. Read the abstracts of all articles from the most recently published journal: are there any that attract your attention? Do the abstracts entice you to read the entire article? Would you reference this article in your research? Does the article appear to have depth, or is the research so superficially explained that it does not appear interesting? These are the same questions being considered by readers reviewing your abstract: write to ensure that your entire work is read and understood by your intended audience.

The following examples of well-written abstracts explain the contents of journal articles and situate the research within the wider field of study.

Garland, David. 'On the concept of moral panic', *Crime, Media, Culture* 4, 1 (Apr. 2008): 9–30.
The article develops a critical analysis of the concept of moral panic and its sociological uses. Arguing that some of the concept's subtlety and power has been lost as the term has become popular, the article foregrounds its Freudian and Durkheimian aspects and explicates the epistemological and ethical issues involved in its use. Contrasting the dynamics of moral panics to the dynamics of culture wars, the author shows that both phenomena involve group relations and status competition, though each displays a characteristically different structure. The piece concludes by situating 'moral panics' within a larger typology of concepts utilized in the sociology of social reaction.

Mccormack, Gavin R., Ester Cerin, Eva Leslie, Lorinne Du Toit, and Neville Owen. 'Objective versus perceived walking distances to destinations: Correspondence and predictive validity', *Environment and Behavior* 40, 3 (May 2008): 401–25.
Judgments concerning features of environments do not always correspond accurately with objective measures of those same features. Moreover, perceived and objectively assessed environmental attributes, including proximity of destinations, may influence walking behavior in different ways. This study compares perceived and objectively assessed distance to several different destinations and examines whether correspondence between objective and

perceived distance is influenced by age, gender, neighborhood walkability, and walking behavior. Distances to most destinations close to home are overestimated, whereas distances to those farther away are underestimated. Perceived and objective distances to certain types of destinations are differentially associated with walking behavior. Perceived environmental attributes do not consistently reflect objectively assessed attributes, and both appear to have differential effects on physical activity behavior.

Safai, Parissa, Jean Harvey, Maurice Lévesque, and Peter Donnelly. 'Sport volunteerism in Canada: Do linguistic groups count?', *International Review for the Sociology of Sport* 42, 4 (Dec. 2007): 425–39.
Given the importance of volunteerism to Canadian sport, and the need for research to understand the characteristics of sport volunteerism, a pilot study was carried out to explore the experiences of volunteers in sport. The study focused on two different sport associations (one individual sport and one team sport) in two majority linguistic localities (one predominantly francophone in Québec and one predominantly anglophone in Ontario). Language is a key element of community membership, and this article pays specific attention to the relationship between language and sport volunteerism in Canada. The results indicate that there are some different patterns of sport volunteering between the two different (official) linguistic communities, and suggest that the experiences of Canadian sport volunteers in relation to linguistic community membership have implications for the recruitment, training, and retention of volunteers in sport, and warrant further research.

While these abstracts may address topics that are not within your field of study, each is interesting, concise, and sharply written. They provide the reader with a clear understanding of the contents of the larger article in few (the longest at 139) words.

Abstracts often end with a listing of keywords in the longer work (stated in alphabetical order). In e-journals, the reader may be able to click on the keyword link to search out other articles that also contain the keywords. Again, give careful consideration to the keyword listing at the end of your abstract to be certain the words accurately reflect the contents of the larger work.

Annotated Bibliographies and Literature Reviews

There are two general methods for categorizing or discussing references in an academic environment: an *annotated bibliography* and a *literature review*. With both types of documents, the student is expected to research written works (books and articles from academic journals, both peer-reviewed and non-peer-reviewed—see Box 3.1), and occasionally non-academic works such as newspaper or reputable magazine articles. They are presented here together, as annotated bibliographies and literature reviews of academic writing are often confused with one another, and it is important to know the difference and to be able to complete either form of writing.

Box 3.1 Peer-Reviewed and Non-Peer-Reviewed Journals

Students will sometimes encounter references that speak to 'peer-reviewed' and 'non-peer-reviewed' articles. Some academic journals will not publish an article until it has undergone peer review. This means that the article is submitted to the journal, then the editor sends the article out to two or more researchers in the same field of study. The researchers review the article carefully to determine if there are any weaknesses in the conceptualization of the problem or the methods used. The reviewers also will look for errors in data analysis or calculations. The reviewers then send their comments back to the journal editor, recommending publication, publication with revisions (which may require another round of the review process), or rejection. Unless the decision is one of outright rejection, the editor passes this information back to the author, asking for required changes. The reviewers are generally not identified to the author to ensure that they can freely comment on the article.

Publication in a peer-reviewed journal is viewed as more difficult and therefore more prestigious for a researcher. Advancement at many universities depends on the researcher's ability to publish extensively in recognized peer-reviewed journals. Most applications for major funding, as well, go through a peer-review process.

Non-peer-reviewed articles are submitted to a publication and included at the discretion of the editor. While the document may be proofed and edited, it does not undergo the rigorous scrutiny of a review by multiple researchers in the same field of study. However, peer reviewing is open to the criticism that it creates and maintains an 'old boys' network', where those who have achieved a certain status in their field determine what is valid and worthwhile within the discipline and what is not. It can be a conservative process that is unwelcoming to new and revolutionary ideas and ways of understanding.

The divide between peer-reviewed and non-peer-reviewed publications is being complicated by the ability of researchers to self-publish on the Internet. Information can be widely dispersed without peer review or any vetting by the wider scholarly community; if the findings are interesting or topical, they may be picked up by the media and given credence by secondary publication in news sources.

Nonetheless, as a student, the validity of research can be partially assessed by the means of publication. If an article is contained in a peer-reviewed, established, and accepted journal, the student can generally place some confidence in the findings. If the research is contained in a non-peer-reviewed journal or is self-published, the student should be more cautious in the tacit acceptance of the findings of the article (although, certainly, good, scientifically valid work is found outside of the peer-reviewed journals). The onus is on the student to be an aware and knowledgeable researcher.

Annotated Bibliography

An annotated bibliography provides a list of references and describes their relevance to the student's research. For example, the following provides two references for use on a research paper on changes in consumer behaviour. The reference is cited, followed by a short description of the text or article:

> Carr, M. 2000. 'Social Capital, Civil Society and Social Transformation', in R.F. Wollard and A.S. Ostry, eds, *Fatal Consumption: Rethinking Sustainable Development*. Vancouver: University of British Columbia Press, 69–97. As stated by the author, the purpose of this article 'is to help us to begin thinking conceptually and broadly about the profound social transformation in both consumption and production that will be necessary to achieve sustainability' (p. 69). The author examines modes of production and consumption, and consumer society's tendency to consider affluence in terms of the possession of material products. The author discusses other models of capital and production and emphasizes the need for wholesale social, economic, and political change needed to effectively address the issue of sustainability.

> Janelle, D.G. 2001. 'Globalization, the Internet economy, and Canada', *The Canadian Geographer* 45, 1: 48–53. Janelle first examines the role of Canadian researchers in advancing the academic understanding of the position of local and regional markets within a global economy. Next, he considers the impact of Internet-based electronic commerce and current estimates of e-business. He concludes with a discussion on the role that the Internet plays in fostering social capital in community life, and notes that the maintenance of socio-economic opportunities at the local and regional level is imperative if social civility is to be maintained in the century ahead.

In an annotated bibliography, the student is sometimes required to add a statement describing the relevance of the text or article to the student's research (that is, a statement that would say 'this article is applicable to my research because . . .'). As with any assignment, be certain of the criteria and expectations of the professor before submitting your work.

Literature Review

A literature review is a section within a larger work or a stand-alone paper that critically analyzes previous academic research in an area of study. It provides context to your research by connecting your project to completed academic works on the same or related topics.

A literature review is more than a short statement on the contents of a reference document; a literature review weaves a narrative that assesses previous works and synthesizes perspectives and ideas into a well-written, concise accounting of what is known or not known in a field of study. A well-written literature review will:

- advance your research from a series of disconnected articles;
- evaluate the validity of previous research;
- identify gaps in the research—areas where further study is required;
- note if previous research appears biased (or worse, flawed or in error);
- establish the strengths and weaknesses of previous research;
- identify opposing viewpoints and unresolved arguments;
- illustrate why your research is relevant and important;
- allow you, the researcher, to build on the work of others and perhaps advance knowledge and ideas within your field of study;
- show that you understand the context of your research—you are familiar with the important works previously published, and are able to understand the context of your work in relation to other research in your field of study.

Box 3.2 Literature Review Formats

Literature reviews can be written in the following formats:

- *Research perspectives.* The literature review develops by grouping the methods (see Part II) or methodologies (the principles by which the methods are deployed and interpreted—the lens of the researcher on the research).
- *Chronologically.* The research is traced through time (usually starting from some key event or seminal work).
- *Gap analysis.* The literature review begins with what is known about a topic, and then identifies gaps in the research (showing why the researcher's work is necessary and important to the field of study).
- *Contrarian.* The researcher sets out historically accepted lines of arguments or methods, then suggests a new approach to addressing the topic or problem. This approach differs from a gap analysis in that the researcher is suggesting that this is more than filling in a gap, it is an entirely new means for addressing issues in the field of study.
- *Inference.* It may be that the research question is so current that there is little published material on the topic. In this case, the researcher constructs a literature review from articles that address similar issues and justifies why these are valid articles from which to draw inferences.
- *Qualitative vs quantitative.* It may be that your research area divided neatly between qualitative and quantitative analyses (for example, research on trends in retailing sometimes focuses on the economics of retailing, and sometimes on the behaviour of shoppers). In this case, the research is divided into these two topic areas, within which the literature may be presented chronologically or from a topic-focused perspective.

It should be noted that a literature review does not have to regurgitate every article or text ever written on a subject—that would be impossible or, at best, cumbersome. Implicit in the literature review is that the student has already reviewed the major works and has selected those that are most relevant or that most clearly illustrate the point being made: the bibliography for the written work will be much more extensive and will illustrate that the student has a broad-based knowledge of the field of study. As a researcher, you must be clear on your topic or problem before developing the literature review so that you know what is relevant and not relevant to your study.

To construct a literature review, the first step is to take as wide a sweep as possible through the existing literature. It may be that a recently published article piqued your interest in the topic; if so, work from the references listed in this article back to the sources that the author considered relevant (then back through the references of the referenced article, and so on, and so on . . .).

Next, review the abstracts or summaries of the texts or articles. Step back and reconsider your research question—does it remain relevant? Has it been done before? Is enough information available on the topic? It may be that a review of the literature changes the research question. Patterns in the research will begin to be observed (such as references to the same seminal text, similarity in research methods or methodologies, or the identification of the same gap in a number of articles). Select those articles or texts that appear most relevant to your research question.

The last step is to read the full articles or texts and begin to develop the literature review. Select a format as outlined in Box 3.2 and group the references accordingly by some common denominator.

Practical tip. It is critically important to develop a system for referencing the works you have reviewed and read: little is more frustrating than remembering a portion of a reference or a quote that would be perfect for your paper and not being able to find it. An efficient method is to use computer programs such as RefWorks or EndNotes to record the references along with a brief description of the work. These programs are available for students through the libraries at most universities and offer a means of recording every reference ever considered by the student. While the data entry can be time-consuming, you will save countless hours in retracing your steps to find that half-remembered reference.

Take advantage of the services offered by your university library to build skills in research. Many universities offer classes, on-line tutorials, how-to documents, or one-on-one assistance in conducting research. A few hours spent early in your university career on learning to use indexes and on-line sources will save many hours over the course of years of research.

The literature review should begin with an introduction that references the research question (but does not fully restate it, as this would have been covered in the introductory paragraphs or section of your document). The format of the literature review should be noted (for example, chronological or perspective-based), along with the rationale for the research (that is, why are you doing this research?). On occasion, this introduction might also identify an area of research that you are *not* investigating because it is either not relevant or beyond the scope of the research question.

The main body of the literature review proceeds to review relevant works. Again, it is not written in point form or presented as a series of disjointed text summaries: this review is the backstory to your research. Weave the articles and texts together in an interesting and descriptive format: the literature review carries the reader through to a conclusion where you prove the importance of your research.

> **Practical tip.** Be cautious of falling into a pattern within the main body of the literature review (as noted in Chapter 2) of 'Jones said that', then 'Smith said that', followed by 'Tom said that'. Instead, use transitions to link the sentences and research by different authors, such as 'Jones said . . . . This was followed by Smith's work, which expressed a slightly different viewpoint. . . . Later, Tom concluded that . . . .' The links help to carry the narrative of the literature review.

> **Another tip.** Be certain to consider the most recent work in your field of study. Recent journal articles are usually more current than book-length studies, but much, of course, depends on the field and the specific research topic. If your subject is historical, for example, the major and important works will have a much longer 'shelf life' than might be the case with a study of recent trends in retailing or of the positive and negative effects of ecotourism in a particular region or country. Just because something is recent does not mean it is of greater value than seminal articles and books in your research area. As you become more conversant in your field and your research area, you will recognize that some of what gets published is of little interest or, for that matter, of little or no value to the greater body of research.

The literature review ends with a conclusion that summarizes the important contributions to the field and how your research fits within this context. As an example, the following excerpt from a journal article by K.G. Jones and M.J. Doucet on big-box stores in the Toronto area[1] provided the following succinct literature review:

> Geographers have had a long tradition of exploring the urban retail structure that dates to the early work of Proudfoot (1937). Over the last sixty years several traditions have emerged, and the geographic literature relating to urban retailing now can be grouped into five research perspectives. The initial area of inquiry focussed on both the identification and classification of various broad structural elements of the retail landscape (Berry 1963; Schell 1964; Davies 1972; Sibley 1976; Potter

1981; Jones 1984). These studies dominated the academic literature until the early 1980s. A second area of interest extended this work by exploring the dynamics of spatial retail change (Simmons 1964, 1966; Shaw 1978). A third research emphasis has been directed toward improving our understanding of the development and operation of particular retail structures—shopping centres, downtowns, retail strips, big boxes, and specialty retail areas (Applebaum 1970; Murphy 1972; Davies and Rogers 1984; Dawson and Lord 1985; Johnson 1991; Simmons 1991; Jones and Doucet 2000). An examination of the institutional and corporate impacts on retail activity has been a fourth area of interest. This tradition continues to explore the role of planning legislation on the development of the retail fabric (Hallsworth et al. 1995; Guy 1996; Hallsworth and Worthington 2000) and the relationship between corporate strategy and retail development (Laulajainen 1987; Doucet et al. 1988; Wrigley 1988; Jones 1991; Smith 1991). Finally, a rich tradition of examining the societal impacts and the implications of institutional power on urban retailing has emerged as an important field of contemporary analysis. This area of inquiry has integrated social and political theory into the discipline's assessment of the complexities of the current retail environment (Hopkins 1991; Jackson 1991; Hallsworth 1992; Mort 1996; Wrigley and Lowe 1996).

In this example, the authors grouped the literature into five research perspectives, providing an overview of relevant research based around areas of inquiry. Also, notice that the researchers grouped the references in brackets after outlining each research perspective. This is a short-form method of ensuring that the reader understands the range of material and authors referenced in the research, and a common format in some academic journals where article length is an issue.

A literature review may also be developed chronologically, tracing previously published research forward from some key point in time through to the present to illustrate the evolution of thought in the field of study. This style of literature review shows a progression or building in the research, ideally reaching an end destination that is your research question. Other methods for formatting a literature review are identified in Box 3.2.

A good method for understanding literature reviews in a field of study is to survey academic journals (most are now on-line and easily accessible through any university's library system). Starting with the most recently published issue of a journal in your academic area, skim through the section titled 'background' or 'literature review' in each article. The review of only a few articles will give you a sense of the expectations of the journal's editors on the breadth and depth expected of a literature review for the publication of an article. Next, seek out lengthier and more detailed reviews in textbooks or dissertations published in your field of study. Note the method of classifying the research (for example, by perspectives or chronologically) and consider the method that would be most applicable to your own research.

For both annotated bibliographies and literature reviews, it is important to note that the first references found are not necessarily the best references. Take time to delve deeply into existing writing on your field of study or interest and understand the methods and biases represented in published works. Not all works should be given equal status in your literature review; be certain that your quoted references are defendable, reliable, and valid.

Term Papers

Every term in almost every course in the social sciences and humanities, students are expected to write term papers (also known as research papers). As with other assigned writing, be certain that you understand the format requirements before initiating any work. That is, are you required to provide an annotated bibliography or a reference list? Is there a specific format? Are there sections that must be included in the paper—a literature review or a compare/contrast section? Clarity on the assignment will ensure that your time is spent on productive research activities that meet the requirements.

At first, the task of writing term papers can seem overwhelming—it can be made easier, however, though adherence to a format for building a term paper.

Establish a Topic

Unless the topic is assigned to you, defining the subject of a research paper may seem extremely difficult. Often, students will select a research area that is too general and become bogged down with too many references and too much material. For example, a sociology student may be interested in studying homelessness in Canadian cities. The student spends seconds on defining this as the topic, and then launches into research for the paper. After spending hours researching through academic journals on-line, the student will recognize the overwhelming number of articles on a wide range of topics, and unless the student's paper is a survey of anything ever written on homelessness, the hours spent have been largely unproductive.

A better approach is to spend some time developing a research question. What specifically are you interested in researching? Is it the role of non-profit organizations in providing services to homeless people? Is it the relationship between federal funding and homelessness? Is it health issues for homeless people? In fact, are you more interested (now that you think about it) in researching shelter use by homeless persons with pets? Narrow the topic to a question that gives shape to the term paper and puts focus to the research.

Research

On-line search engines for academic journals (accessed through university libraries) are a good place to start to research a topic for a term paper. Use keyword searches to begin to build a reference base. Once you find an article of particular relevance, review the keyword list if one is included with the article, then use these terms to advance your search. In addition, review the references attached to the article and use these references to branch out into new areas. Next, seek out academic textbooks on the topic, tracing back through key authors identified in the academic journals. It is likely that prolific writers will have published in both journal and textbook formats.

It is important to note that the first references you come across are often not the best references. Allow yourself ample time to fully research a topic. You will know your research is complete when the references become repetitious or circular—the references you already have are referencing other references you already have.

Other sources are news and magazine articles and websites. While these may not be peer-reviewed academic sources, they may provide useful and current information on the topic that will assist in focusing the research question.

For all sources, consider the perspective of the writing. Is the author attached to a research institute known to have a particular bias? Is the article published in a journal that only considers works that match the opinions of the editor? Is it research-based (with methods clearly identified) or an opinion article? Be a critical reader and note if the referenced article appears less than objective.

Be certain to note the bibliographic information for each reference as you work, either using the on-line programs previously noted or your own referencing system. Some students use a notecard for each reference that also includes a short statement on the applicability of the reference to the research. Others construct databases that allow for keyword searches. Whatever the method, just be certain to use one.

Build an Outline

Using either a linear or circular method (or a combination of both), build an outline for the paper. Refer back to the assignment and be certain all necessary sections are noted in the outline. Seek out a logical order for the progression of the paper—start with an introduction that states the research question, look for logical groupings in the research material that will form different sections in the paper, and end with a strong conclusion that restates your arguments and the purpose of the paper.

Build a Draft

Using your references, start to build on the outline as you write. Focus on the relationships among the references. Add footnotes (either at the bottom of the page or at the end of the paper, depending on the preferred format) that offer additional information as well as referencing articles. Add charts and illustrations, if they are applicable and add meaning to the paper.

Once a draft is largely completed, consider if the paper retains logical order or if sections should be reorganized to clarify meaning and improve the flow of the paper. Review the references one final time to ensure that all relevant references have been included and correctly cited.

Final Version

After writing the draft, if possible, put it aside for 24 hours. The draft can then be reviewed from a fresh perspective. Critically read what you have written, looking for gaps in the information presented or errors in referencing. Look for spelling and grammatical mistakes and places where sentence structure can be improved to maximize readability. Consider the flow of the paper: Is it logical? Does it read well? Finally, review the research question: Have you answered it?

Make any necessary changes to the text, review once more for formatting and style consistency, such as indented quotations and the use of quotation marks and italic, then again to ensure all technical requirements for length, margins, and format have been met.

While it is not always easy to do, be sure to allow yourself time to develop the research question and work through the draft of the paper. Using this approach will assist in maintaining focus, will save time, and will result in a higher-quality research paper.

Research Proposals

A research proposal is a specific form of academic writing prepared for one of two reasons:

- to gain approval for research;
- to gain funding for proposed research.

For the former, the research proposal outlines the research questions and methods that will be used, considers ethical issues, and presents anticipated conclusions to the research. The proposal may also set out a time frame for completing the research. In most cases, the proposal will also include a literature review that supports the research question and verifies that the research is relevant.

These research proposals are typically written to gain approval from an organization that will allow the researcher to advance through an academic program. For example, most advanced degrees require that the student write a research proposal that is approved by a committee of professors with expertise in the student's area of study. The student's proposal forms the basis of the thesis or dissertation and sets out the program of study.

For the latter, all these factors are considered along with a proposal for funding. The costs of the research are developed in detail, and a funding organization then evaluates the research proposal against all other applications also requesting research funding. Along with proving the validity and relevance of the research, the researcher must show why his or her research is of particular value and deserving of funding.

For both types of proposals, the document must address:

- What do you propose to do?
- How do you propose to do it?
- Why is the question relevant?
- What will your research accomplish?

Most organizations will provide a required format for a proposal, and most will contain the following:

- *Abstract*. As discussed at the beginning of this chapter, an abstract provides the research question, proposed methods, and anticipated results. While an abstract for an academic article will speak to work already completed, the abstract in a research proposal identifies what the researcher intends to do, should the proposal be approved.
- *Introduction*. The introduction states the research question and verifies why the question is important. The question might be stated in the form of a hypothesis (see Part II) or as a more general inquiry. The context of the research may be established (how the research question relates to previously conducted research), along with sub-areas of investigation. In addition, the researcher may identify areas that will not be reviewed, as they are either irrelevant to the research question or beyond the scope of the research.
- *Literature review*. As discussed above, the literature review establishes that the research is worthy of academic study by showing how the proposal relates to previous research. It is important to develop conceptual links among reputable academic works and your own research: these links might address how your research builds on the work of others or show that your research will fill a gap in your field of study.
- *Methods*. In this section, the writer outlines how he or she will address the research question. The methods might include the development of a survey, fieldwork, archival research, interviews, experiments, and computer or lab-based analysis. Necessary equipment (for example, measurement instruments or recording equipment needed for fieldwork) is identified, along with a proposed method for obtaining these supplies. This section will identify the test subjects for the research, if it is to involve human or animal subjects, and if approval from an Ethics Review Committee[2] is required. A timeline may be included to document how the researcher intends to work through a program of study, research, and writing.
- *Conclusion*. The final section of the proposal will identify the anticipated results of the research, although it is recognized that the results may change substantially after the research is completed. Any anticipated issues are also identified, such as concerns over the cost of the research or the availability of necessary resources (such as lab time).
- *Bibliography*. The researcher likely will need to produce an annotated bibliography to verify that the relevant references for the study have been searched out. Be certain to cite all references in the required format without errors or omissions.

Finally, do not underestimate the importance of a good title: an explanatory and interesting title will attract the attention of the reviewers and will help to give clarity to the research topic.

When you prepare a research proposal (either for approval or funding), the most important issue is to follow the required format. Often, the reviewing organization will refuse to consider any proposal that does not meet the criteria as outlined: it is disappointing (or worse, research-ending) to know that your application or proposal was refused because you did not meet a small technical requirement.

A research proposal needs to be not only technically accurate but also engaging and interesting. You are requesting approval or funding, and the reviewer needs to be certain that your proposal has merit. Write to make it easy for the reviewer to sponsor your research.

Theses and Dissertations

As applied to a lengthy written work of research, the terms 'thesis' and 'dissertation' are sometimes used interchangeably. Generally, however, a *thesis* is a major research document required to finish a master's degree program, while a *dissertation* is required for a Ph.D. and may include additional criteria that evaluate the research as either original or substantial (or both).

The key element of a thesis or dissertation is the hypothesis or research question. Give yourself ample opportunity to develop a question that is relevant, important, and interesting. The latter is most important: you will be spending years focused on this question. Be certain it is something you find captivating, a question that will spur you through the lengthy time spent researching and writing.

It can be useful at the outset to read through a number of theses and dissertations to get a sense of the depth of research and level of discussion required for these documents. This review will help you develop a picture of the final format of your own research as well as aid in focusing your research. Ideally, a review of a series of well-written and interesting documents will clarify any questions you may have on developing a thesis or dissertation. Bear in mind that, essentially, the thesis or dissertation is a book-length study of a particular limited topic, and some eventually become published books.

Once you have a sense of your research question, you will work with your advisor to develop a research proposal. This proposal will be vetted through a review committee, and you may also be using the approved proposal to seek out funding sources that hopefully will sponsor both your research and your tuition and expenses while in graduate studies. This proposal will then form the basis of your thesis or dissertation, as it will include an introductory chapter that outlines your research topic, a literature review that situates your research (although you may continue to add new references to the literature review as you work through your program of study), and your research methods. The research proposal sets out what you are going to do, and the completion of the thesis or dissertation is the doing of it. Therefore, the format for these latter documents is similar to that of a research proposal. The key differences are that the research is now complete, and the completed document clearly illustrates the advanced knowledge of the researcher and also contributes new or more detailed work to the subject area.

Most universities publish detailed guides, outlining the expectations for theses and dissertations. For example, the Graduate Studies Faculty at the University of Victoria provides a guide to students describing the format for title and abstract pages, copyright licences, and committee approvals. The document also provides a 30-point checklist for formatting, including margin requirements and the placement of page numbers. Format is important. Be certain you are aware of the smallest requirements and work to meet them. It is easier to do the formatting correctly from the outset than to go back and amend your work.

Chapter Review

This chapter reviewed likely formats of writing that will be encountered in an academic environment. The most important issue is to follow the required format: in academic writing, the author is generally seeking a grade or approval, or perhaps funding for a research proposal. Losing marks or funding because the document did not meet the required format is both disappointing and unnecessary, and below the standard expected at a university level.

Review Questions and Activities

1. Explore funding organizations for academic research in your field, and compare the application criteria among these organizations. What are the key elements that must be contained in a funding proposal?

2. A good literature review follows a logical format. What are the generally accepted formats for literature reviews? What is the most commonly used format?

3. What is the difference between a term paper and a thesis or dissertation?

4. Go to <www.collectionscanada.gc.ca/thesescanada/index-e.html>. This is the portal for Theses Canada, an on-line catalogue of full-text electronic versions of theses published across Canada. Enter in search terms that you are interested in, and review current research in your area of study.

5. Find out about your university's Ethics Committee. What are the requirements for conducting research on human subjects? On animals? Are there requirements for other kinds of research?

Chapter Four

Writing Formats for Professionals

In your academic career, much of your written work is in the form of essays or term papers. You begin with a question (or set of questions), find references in that field of inquiry, obtain data, complete an analysis, and draw conclusions. Most academic papers follow a standard format that is useful for these large-scale, comprehensive research works, but this format is less applicable to professional writing situations. In a professional work environment, you need to produce information in a variety of formats, each with its own purpose and criteria. This chapter provides tips to improve your ability to write professionally and outlines nine different formats commonly used in professional writing:

- press releases,
- newsletters,
- briefing notes,
- proposals,
- internal memos,
- business letters,
- e-mails,
- resumes,
- cover letters.

Press Releases

A press release is a statement prepared for distribution to the media (print, television, radio, and on-line) intended to make an announcement or provide up-to-date information that is useful and interesting. These are also referred to as 'media releases', in recognition of their wider distribution among a variety of media outlets. In addition, the press release may be used to invite the media to an event—if this is the case, a follow-up may be necessary to confirm their attendance.

A press release is a brief document—no more than one or two pages. It contains a series of short paragraphs that provide sufficient information to pique interest, but typically the press release does not tell the entire story.

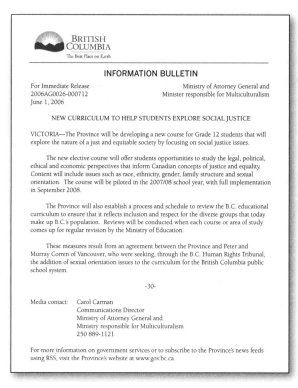

Figure 4.1 Provincial Press Release

Often, a press release is intended to garner interest from the editor, who will then assign a reporter to follow up on the press release (ideally with a lengthy feature article). As well, press releases can be used to make an announcement (such as the date of an event). In this instance, the press may publish the press release 'as is', then follow up by attending the event. A press release may also be a factual announcement (identifying the new CEO of an organization, for example, or the results of new research conducted by or on behalf of an organization). Again, the release may be used by the media as is or trigger a longer, more involved story. Ultimately, any use of the press release by a media outlet is a measure of your success.

While variation is available in the design of press releases, the following outlines an acceptable conventional format to ensure greater likelihood that your press release will be read by a decision-maker at the media organization and picked up for distribution.

Top of page: letterhead or logo of sponsoring organization.

Under letterhead/logo: Write 'press release' (or a similar descriptive term, such as 'news release' or 'information bulletin') at the top of the page. This informs the reader as to the purpose of the document.

First line: Write 'FOR IMMEDIATE RELEASE' in bold and all capital letters. This informs the media outlet that the information is date sensitive and current.

Second line: Leave a blank line or space.

Third line: Write your headline in a combination of capitals and lower-case letters in bold. The headline should be just a few words that capture the meaning of the article. Write a headline that relates to the subject matter and captures the reader's attention. Do not use a pun or word play, unless you are certain that the headline is amusing and inoffensive. You may want to test the headline with your colleagues to see if they agree that it accurately captures the spirit and tone of the article.

Fourth line: Leave a blank line or space.

Fifth line: Write the date and place (city, province) where the press release is written. This provides the when and where of the press release (see Chapter 7, Research Basics).

First paragraph: This paragraph contains all relevant information on the who, what, where, when, why, and how of the issue in one to three sentences. This paragraph should be able to stand alone: if the press reports on no more than this first paragraph (possibly including it in a 'news shorts' column or in the running text at the bottom of the television screen), the public should be able to understand the purpose of the press release.

Middle paragraphs: The paragraphs following the introductory paragraph add further information in an 'inverted pyramid' format. This means that the most relevant information is presented first, then followed with background information in a reverse hierarchy from most important to least important. These paragraphs provide additional detail to flesh out the information. Each paragraph contains information that can stand alone, and the media outlet that chooses to use the press release could select only the first paragraph or add some of the middle paragraphs to create the news story. At least one of the middle paragraphs will contain a quote from a source or authority to add credibility to the press release and offer a 'human voice' on the subject. Make sure that you have permission from the individual to use the quote. If you need to go to a second page, write 'continued' at the bottom of page 1.

Final paragraph: The last paragraph provides background on the individual or organization sending the press release.

Close: The symbol -30- or ### is used at the end to indicate that the press release is complete. This is a convention that began with the teletype machine to show that a transmission had been completed.

Contact information: Careful attention must be paid to identifying a contact person. In some organizations, all media contact is assigned to a person or department, and all communication is channelled through one source. In others, the contact person is the writer of the press release or the individual with the greatest knowledge on the topic. If it is anticipated that the media may have further questions on the press release topic, it makes sense to ensure that the 'expert' can be easily accessed. Still others use the highest-ranking individual at the organization as the contact. While this can add credibility to the press release, it is not recommended if the person is not available to provide further comment. The person must be available once the press release is sent to the media outlets. A telephone number provides the most immediate contact, while an address may be provided for future reference. Do not use an e-mail address as a contact unless the contact person will be monitoring his or her mailbox. A further concern with e-mail is that a response is provided in writing, yet the writer may not take the time to ensure that the information is accurate and well presented. Remember: the more personal communication is, the more likely it is to convince and to be understood. Face-to-face verbal communication is best, followed by telephone communication, and, lastly, by written communication, of which e-mail today is the least formal and persuasive.

10 Rules for Press Releases

1. **Be newsworthy.** Major media outlets receive thousands of press releases every year. What is so special about your event or issue? Does it warrant a press release? If you are looking for a media outlet to do a feature story on your organization, a better approach would be to contact a reporter or editor, explain your issue, and invite her or him to interview people within your organization. A press release is best used to make an announcement or when you have something interesting and timely to present to the public.

2. **Be interesting.** The editor or reporter will likely only read your headline and first paragraph before making a decision on your press release. Make yours interesting and informative—write a press release worthy of being published. Present the who, what, where, why, and how of the issue in the first paragraph so the editor or reporter understands the issue.

3. **Research the media outlets.** Do your research on the media outlets. What content seems more likely to be published by a particular outlet? Are there two competing media outlets in your target area, and is one more credible? It may be that certain outlets publish only certain types of information—consider what you might see covered in a supermarket tabloid compared to a national newspaper. The tabloid is much more likely to publish your announcement on a sasquatch sighting while the national paper might be interested in your research findings on a recently completed study. Instead of a shotgun approach, send your press release only to media outlets that you find credible or at least likely to publish it.

4. **Be economical and to the point.** A press release is not an essay. Make every word count. Keep every sentence to the point. Use clear language written at the level of the target audience. And by all means, bear in mind that what a paragraph is in the popular media is a rather different animal from what it is in a journal article or research paper. In the latter, your paragraph develops fully a single aspect of your topic or main idea in a coherent manner; in a news story in the local paper, a paragraph is a visual sound bite, often of only one sentence, rarely of more than three or four short sentences. This has to do with the newspaper format of narrow columns, so that if paragraphs were fully developed, they would fill entire columns and be difficult to digest for the reader. Also, except for the op-ed pages of the paper, readers do not expect or want extensive analysis—just the facts.

5. **Use the correct format.** The format noted above provides an outline for a press release, but certainly flexibility is permitted. The 2010 Vancouver Olympics Committee sometimes uses '-2010-' instead of '-30-' on press releases, which cleverly emphasizes the date of the event. The communications professionals are aware of the convention of -30-, but have chosen to break the rules to create emphasis (know the rules to break them!).

   Web-based press releases are often more colourful and contain photographs that add to interest. However, an editor or reporter may be less likely to read an e-mailed press release, negating the benefits that might have been gained from photos and colour. Whether paper-based or web-based, a standard format allows the reviewing editor or reporter to quickly understand the purpose of the press release, although creative variances will allow your press release to stand out from the rest.

6. **Be time-sensitive.** A press release gives information on something current and relevant. Obviously, daily newspapers are looking for new information for the paper each day and reporters must work to tight deadlines. Papers that publish with a bi-weekly or longer time frame also work to deadlines. Some on-line news sources are constantly updated, while others are revised on a schedule. Confirm deadlines in advance to ensure your press release matches the publication schedule.

7. **Get a contact.** Your press release is most likely to be used if you have direct contact with either a reporter or editor at the media outlet. Do some research. Is there a reporter that writes frequently on issues related to your topic? Does the newspaper list the editors and reporters and their areas of specialization? If possible, hand-deliver your press release to the person most likely to be interested in it.

8. **Be right.** All spelling and grammar must be checked and rechecked. Be certain your press release is well-written.

9. **Ensure accuracy.** Every fact in the press release must be verified before being distributed to the media. At best, a mistake or inaccuracy will be embarrassing. At worst, it could have long-term ramifications for your career and organization. Circulate your press release to knowledgeable colleagues for review before sending.

DEPARTURE BAY NEIGHBOURHOOD PLAN – COMMUNITY SOLUTIONS WORKSHOP

Do you want to have a say in the future of Departure Bay? Do you have ideas or proposals on
how the community's Neighbourhood Plan could help address issues related to development &
redevelopment, the environment, traffic, parking, safety, or recreation?

If so, then come to the Departure Bay Neighbourhood Plan Community Solutions Workshop! The
workshop will build on the issues identified in October's well-attended Departure Bay
Neighbourhood Plan Open House. The Workshop will be a great opportunity for community
members to help shape Departure Bay's Neighbourhood Plan.

The Community Solutions Workshop is part of a series of public events hosted by the City of
Nanaimo, the multi-stakeholder Departure Bay Neighbourhood Plan Working Group, and In
Vision Planning, a group of graduate Students from the University of British Columbia. All
residents of Nanaimo are invited to attend. Snacks and refreshments will be served. Childcare will
be available.

When: Wednesday, December 7, 2005. Doors Open at 6:00 p.m. Workshop starts at 6:30 sharp,
and will conclude by 9:30 p.m.

Where: Departure Bay Activity Centre, 1413 Wingrove Street.

For more information, please contact: invisionplanning@gmail.com or (250) 755-4483

Figure 4.2 Municipal Press Release

10. **Use attachments only when necessary.** If necessary, a separate 'press kit'
    may be provided (either on paper or electronically) along with the press
    release. If the reporter or editor is interested in further information, he or
    she will contact you. A website address may be included to direct the media
    outlet to additional information. If a website is referenced, be certain that the
    text is up to date, that the links included remain operable, and that the pho-
    tos are ones that you would consider for publication by the media outlet. The
    reporter or editor may download information directly from your website, so
    be certain it reflects the current state of your organization.

What Not to Do in a Press Release

- Don't e-mail your press release to a general contact mailbox. Your press
  release will likely never be read and will be deleted as spam.
- Don't ignore conventions. If you want your press release to be picked up
  by the media, stick to the conventions in format so the reporter or editor
  will not have to search for the information. Ensure the press release is not

overcrowded or illogically organized. Also, including spelling mistakes and grammatical errors to further emphasize that your press release is different from those that the media outlet normally publishes will guarantee that it goes straight to the bottom of the circular file.

- Don't use jargon. A press release that is rife with acronyms or technical terminology will not hold the reader's interest.
- Don't write an advertisement. The press release is a means of providing information, not selling a product or service.

Newsletters

Newsletters are a useful way to send a message to a target audience. They differ from press releases in that the writer or organization retains complete control over the content of the newsletter, and also over the distribution and timing. If a series of newsletters is published, the target audience can gain familiarity with the format and content of the newsletter and a relationship will be built between the newsletter publisher and the target audience.

The first question to consider when producing a newsletter is commitment—there are both time and dollar costs to the design and delivery of a newsletter.

Box 4.1 Advantages and Disadvantages of Newsletters

Issue	Advantages	Disadvantages
Costs	• Relatively inexpensive compared to print or television advertising. • Mail costs can be reduced if the newsletter is sent as bulk mail or business mail. • Costs can be further reduced with a web-based newsletter.	• Costs can escalate if a great deal of time is spent designing and writing the newsletter. • Higher costs for direct mailing. • Time requirements for managing a mailing list.
Content	• Full control over content.	• Difficult to confirm if newsletter is being read.
Timing	• Full control over timing.	• If published to a schedule, time frames must be met.
Readership	• A regular newsletter is a good way to create a relationship with a target audience.	• People are increasingly refusing bulk mail: the newsletter may not reach the recipient.

News from NALT
Newsletter of the Nanaimo & Area Land Trust Society
July, 2008

MOUNT BENSON GOAL IS REACHED!

Thanks to a grant of $50,000 from the Mountain Equipment Co-op, NALT has now reached its goal of raising $475,000, half the purchase price of the 523-acre property on the upper front face of Mount Benson. The other half was provided by the Regional District of Nanaimo, which will hold the property as a Regional Park.

"We were thrilled to receive the news of this generous grant from MEC!" said NALT Board Director, Dean Gaudry. This donation is very timely for NALT. It carried the Mount Benson fundraising over the top, and completes more than two years of a very visible campaign that has garnered such strong support from residents and businesses throughout the Nanaimo area.

Apart from the MEC grant, every penny of the money raised for Mount Benson has come from the community. More than 1400 local donors and sponsors have contributed to the purchase of the mountain in small and large ways – from cash donations of less than $20 to more than $95,000 in shares donated by one very dedicated Mount Benson supporter.

NALT is calling this "Phase I" of the Mount Benson Campaign. There have already been informal discussions with other property owners on the mountain, and NALT is poised to begin Phase II. With that in mind, NALT will continue to hold fundraisers and accept donations for the mountain. However, Phase II will not entail another high-p... properties that are negotiated will be purchased... sums to be raised annually.

Figure 4.3 Example of a Paper Newsletter

Note the masthead, table of contents, contact column, and clear graphics.

News from NALT

is published by the Nanaimo & Area Land Trust

NALT's Mission is:

to promote and protect the natural value of land in the Nanaimo area

The Nanaimo & Area Land Trust Society was registered as a B.C. Society in 1995, and subsequently was granted charitable tax status and the right to hold conservation covenants.

Executive Director: Gail Adrienne
Executive Assistant: Austen Scott
Plant Nursery Manager:
Susan Fisher
JCP Crew Coordinators:
Susan Fisher and Austen Scott
Administrative Assistant:
Betty Penston
Financial Manager:
Deanna Bickerton
Volunteer Coordinator:
Paul Chapman

Board of Directors:

Holly Blackburn
Gillian Butler
Dean Gaudry (co-chair)
Barbara Hourston (co-chair)
Dale Lovick (recording secretary)
Lynn Peterson (treasurer)
Shelley Serebrin
Ron Tanasichuk
Jim Young

Newsletter Editor: Ken Lyall

Contact us at:

The NALT Stewardship Centre
Madrona Building (lower floor)
Suite 8, 140 Wallace Street
Nanaimo, B.C. V9R 5B1

(250) 714-1990
admin@nalt.bc.ca

www.nalt.bc.ca

RALPH HUTCHINSON, 1930–2008

It was a sad day on March 20th when the news came that Ralph Hutchinson had passed away, following a several years' battle with cancer. Rafe was a valued member of the NALT Board for almost five years.

NALT first met up with Rafe in 2002, when, as a recently retired B.C. Supreme Court judge, he chaired the consortium of property owners of the 145-acre property in the Linley Valley that NALT was working to acquire. Thanks in great part to Rafe's keen negotiating skills, both with the group of property owners and with the City, NALT was able to complete the Linley Valley campaign successfully by the deadline of June 30th, 2003.

Rafe's positive working relationship with NALT throughout the final year of that campaign led him to join the NALT Board in September, 2003. By that time, NALT was in the early stages of communications with the owners of the 523-acre property on the face of Mount Benson, and Rafe, as a long-time mountain climber who knew Mount Benson well, added his considerable skills to these negotiations.

Rafe's last initiative on behalf of NALT and the Mount Benson campaign was to organize an evening of fine dining at the Drift Restaurant on January 24th this year, which raised $4300 for the campaign.

We regret that Rafe didn't live to share our celebrations of the successful completion of the Mt. Benson campaign. We would like to think, however, that he knows we made it to the top.

Rafe on Mount Benson in 2003

Our thanks to more than 1,400 local donors and sponsors who gave so generously to ensure that Mount Benson will remain green forever! NALT staff are working to complete the task of compiling an updated list of all your names, listed in the category of your donation amount. Look for your name on a full page ad in the local newspapers soon.

News from NALT   July 2008 Page 2

Time Costs

How much time do you have to devote to designing the newsletter and publishing each issue? Be realistic in your time estimate. The development of a logo and format can be time-consuming, and a regular publication schedule means that you will need to spend time to publish each issue.

Budget Costs

What is your budget? Your budget amount will impact the number of publications per year, the production quality, and the publication form of the newsletter. With a printed newsletter, costs increase with the use of colour, specialty paper, and increased number of pages. With a web-based newsletter, costs increase with time spent designing and updating the newsletter. Consider both the cost of each issue and costs over the lifespan of your newsletter. If your intention is to build a regular readership, can you afford to keep publishing the newsletter?

On a very few occasions, a targeted readership may be willing to pay for a newsletter. While this may be your ultimate goal, it is likely that many issues will be published before a sufficient and motivated reader base is established.

There is also the possibility of selling advertising in your newsletter, but again, any business person will want confirmation that the investment is worth the cost. You will be required to have readership statistics and potentially a demographic profile of your readership to attract advertisers—this information may only be available after your newsletter has achieved some success and budget can be spent on researching the readership.

How to Write a Newsletter

- *Create a masthead.* Place your logo (or create a new logo) along the top or side of the first page to clearly identify your newsletter. This masthead may be the same as the letterhead used by your organization. Ideally, it is a signature item that the readership will use to quickly and positively identify your newsletter.
- *Include identifiers.* List the name of your organization, the date, the issue number, and the title of the newsletter either as part of or beside the masthead. These identifiers tell the reader who is publishing the newsletter (again, speaking to credibility) and may encourage them to seek out back issues.
- *Create a boilerplate.* The boilerplate is the format of the newsletter that will be used each time you publish an issue. The boilerplate creates a place for each item in the newsletter (table of contents, feature story, graphics, and contact information). Over time, your readers will become familiar with the format of the newsletter and will be able to flip quickly to familiar sites. The use of a boilerplate also makes it easier to produce the newsletter. While the content of the articles changes with each issue, their placement remains roughly the same, allowing the writer to estimate the required number of works and depth of content needed for each page in the newsletter.

Figure 4.4 A Web Newsletter

Note the links to other sites—each one is active. A web-based newsletter can be more colourful for the same cost.

- *Choose a colour scheme.* If you can afford to add colour to a printed newsletter, do so. Colour makes a newsletter stand out from all the other mailings, and use of a consistent theme can make your newsletter easily recognizable: even adding one colour when printing can make a difference in the appearance of the newsletter. At the same time, the overuse of colour can decrease the professionalism of your newsletter. Look for a balance that creates interest but maintains coherence.

- *Create a personality.* The boilerplate will also set out the 'personality' and tone of your newsletter. Consider the format of a favourite newspaper or magazine—what is it about the format that you find appealing? Consider the look of publications that you would never read or buy. Is it the format or content (or both) that is distasteful? Your reader will form a first impression of your newsletter within seconds of seeing it—be sure that you present your best 'face' for the newsletter.

- *Choose a font.* Fonts with serifs (the small extensions on ends of letters seen in fonts like **Times New Roman**) appear more formal than sans serif fonts (like **Tahoma**), but sans serif tends to be more difficult to read. Often, newsletters use one font for titles and another for text. The font used in your masthead may also be different than the font used in the body of the newsletter. Avoid using overly complex fonts as they can be difficult to read, and also refrain from the overuse of character or 'fun' fonts, as your newsletter will lose professionalism.

- *Add a table of contents.* If your newsletter will be longer than a single sheet of paper, it is helpful to list the contents of the newsletter to alert the reader to important news. Your newsletter may contain feature articles, news updates, contact information, profiles, announcements, or a question-and-answer section. The amount of content will depend on the length of your newsletter.

It is not necessary to fill in every space in the newsletter with content. White space (sometimes referred to as 'negative space', with 'positive space' being the text, photos, or illustrations) helps to lead the reader's eye to key features and gives shape to the newsletter. A three-column format with some graphic appeal and white space is more likely to be read than a solid page of 10-point text. Do not overcrowd the newsletter—add a new page if necessary (knowing that this will increase your costs).

In addition, each new article should be marked with a headline so the reader can easily navigate through the newsletter. If possible, keep articles complete. When formatting or boilerplate requirements cause you to split an article between two pages, clearly label the end of the first section with 'continued on page 2' and the beginning of the second section with 'continued from page 1'.

Finally, carefully consider the use of humour in your newsletter. In an increasingly smaller world, it is becoming more difficult to predict that your humour will be correctly interpreted or that all readers will find a particular joke amusing. If an item may be misconstrued, test it against several colleagues to see if they understand your intent.

- *Reprints.* A newsletter may also contain a reprint of an article that has been published elsewhere. If the article is by a recognized expert, this can give your newsletter credibility. Be sure to obtain permission from the author or publisher before reprinting an article.
- *Illustrations.* Relevant photos, maps, or illustrations can be useful to readers and add interest to your newsletter, although most will require a significant amount of space. Be cautious about the use of public domain clip art (see Chapter 14)—because clip art is free and widely accessible, you may not be the only organization using it, therefore limiting its effectiveness as a 'brand' for your newsletter. Often, a simple, clean format is preferred to one crowded with unnecessary ornamentation or random clip art. If budget is available, you may also consider purchasing stock photos or graphics. Again, the exclusive use of these images is not always possible, and the generality of the subject matter may add little to your newsletter.

    If you use a graphic or photo, be sure to include a caption explaining what it is. Readers are interested in the names of people in photos, locations, or the point of a diagram. Let them know what they are seeing.
- *Time-sensitivity.* Ensure that your newsletter will reach the readers before any time-sensitive information becomes dated. If you are advertising a meeting or event, it makes little sense to include it in your newsletter if the reader will not be notified until after the event is over.
- *Interactive component.* Do you want the readers to be able to provide feedback on your newsletter? Will you publish their responses? Providing a space for a response or contact information is one way to build your mailing list, allowing you to target interested readers. However, a good mailing list requires some maintenance. You may be constantly adding interested readers, but will also need to delete those who request to be removed from the list. This is particularly important for a web-based e-mail, as the ease of responding to an interactive component means you will have more people seeking both inclusion and exclusion from your mailing list.
- *Contact information.* Even if your newsletter is hard copy rather than disseminated through cyberspace, a web address can be useful for people seeking to contact your organization. For both web-based and paper newsletters, only include contact information that is regularly monitored. If you will not respond to phone messages, do not list a phone number. Only include an e-mail address if it is frequently monitored. Decide on the scope and extent of direct contact you intend to have with newsletter respondents before publishing your first newsletter. A successful newsletter (or a controversial one) may generate a barrage of unexpected contact that will take time and budget to manage.
- *Web-based newsletters sent in an e-mail to readers.* While the costs may be substantially less for an e-mail newsletter, there are other issues to consider. It is easy to delete an e-mail notifying of the availability of a newsletter, or to scroll past the notice. If the newsletter is the text of the e-mail, the newsletter may not appear to be important or authoritative. The ease of production should be weighed against these costs to credibility.

- *Web links.* If web links are included, either in the body of the e-mail or in your newsletter, make sure they work. It is frustrating to click on links that either never finish loading or have been broken.

As with any other form of professional writing, it is critically important to check your spelling and grammar. Leave yourself adequate time to carefully proofread your publication before sending it to your readers. Consider the costs if errors are found—although you can publish a retraction in your next newsletter, both minor and serious mistakes make the newsletter appear less reliable and authoritative.

Briefing Notes

Decision-makers are required to keep track of many often unrelated issues. In government or large organizations, leaders may not be involved in the day-to-day issues on a project, but they are required to speak to it and must appear to fully understand the issues. A briefing note is a short, direct communication that provides information to a decision-maker on a subject or issue.[1] Often, a manager or supervisor will solicit a briefing note from a staff person known to have knowledge in an area or the ability to research a question (in increasingly lean organizations, a good researcher who has the ability to apply research skills to any question is highly valuable—this is discussed further in Part II). Sometimes, a staff person will draft a briefing note as a means of formally informing the manager or supervisor of an evolving issue. A good briefing note is clear, concise, and written in plain language. This form of communication is short—often an organization will establish a one- or two-page maximum for the briefing note, with the opportunity to attach appendices if required. Every word in the briefing note is important and there must not be any opportunity for misinterpretation. If technical jargon or acronyms are used, they must be fully explained (with the inclusion of a phonetic pronunciation guide if there is an opportunity to mispronounce a word).

While there are many variations on the format of briefing notes among organizations, the key elements remain the same:

- *Purpose.* State the issue or topic in one or two lines. Why is this briefing note important?
- *Summary of the facts.* This section provides necessary background information. The who, what, where, when, and why of the situation are described. If there are previous briefing notes on the topic, they can be referenced in this section. In non-judgemental, technical language, this part of the briefing note explains only the issues pertinent to the current discussion.
- *Current situation.* The briefing note may contain a section titled 'current situation', which details why the note is being brought forward today. What is happening right now? Is there a deadline to making a decision?
- *Key information.* In addition to the summary, is there any other information that must be provided to the decision-maker?
- *Options/alternatives.* What are the options? What will happen if one option is selected over another? What will happen if a deadline is missed? If more information is required, additional details may be attached as appendices.

Briefing Note

City of Calgary 2006 Annual Water Quality Report

Description

On June 10, 2007, The City of Calgary will release its 2006 Annual Water Quality Report. It will be delivered to more than 300,000 households through the Calgary Sun, as well as by Canada Post. This the fourth year Calgarians will receive the report about the quality of their drinking water. Calgarians can also access the 2006 report on The City's web site at www.calgary.ca or by calling 311.

Key Contact

Public Inquiries To:
311 or
ccweb@calgary.ca

Media Contact: **Internal Contact:**

Current Status / Next Steps

- The City of Calgary, Water Services will release its 2006 Annual Water Quality Report to more than 300,000 households, beginning June 10, 2007.
- The four-page report will be delivered through the Calgary Sun and Canada Post.
- The Water Quality Report will also available at www.calgary.ca or by calling 3-1-1.

Key Messages

- The City of Calgary continues to provide Calgarians with safe drinking water that meets or exceeds all federal health-related guidelines and Alberta Environment water quality standards. In 2006, The City tested more than 109,000 water samples to ensure those standards and guidelines were met.

- The City of Calgary believes all Calgarians have a right to know about the quality of their drinking water and that it has a responsibility to provide this information to the public. Water utilities in the United States are required by the Environmental Protection Agency to publish an annual water quality report for their customers. Although not required in Canada, The City of Calgary, Water Services has chosen to provide this information to its customers.

- The 2006 Annual Water Quality Report also provides information about:
  - what The City is doing to protect our waterways,
  - how Calgarians can conserve water,
  - the quality of water going back into the river after being treated by the wastewater treatment plants, and
  - storm water information.

Frequently Asked Questions

1. Why does The City of Calgary issue an annual Water Quality Report?
The City of Calgary, Water Services believes all Calgarians have a right to know about the quality of their drinking water and that The City has a responsibility to provide this information to the public.

2. What is the quality of Calgary's drinking water?
Calgary's drinking water remains of high quality and continues to exceed all federal health-related guidelines and Alberta Environment water quality standards.

Figure 4.5 Example of a Briefing Note

Briefing Note

City of Calgary 2006 Annual Water Quality Report

3. How does Calgary's drinking water compare with other cities?
The City of Calgary monitors approximately 150 different parameters making it difficult to compare each parameter with other municipalities. When comparing Calgary's overall water quality with other communities, Calgary scores high.

4. Has the water quality changed since last year?
There are always minor variations in Calgary's drinking water from year to year. However, due to The City's treatment process, the quality of Calgary's drinking water remains consistent high, despite occasional (and sometimes dramatic) changes in the quality of our raw (source) water supply.

5. What determines good water quality?
Although consumers often only attribute water quality to taste and odour, The City of Calgary monitors our water for approximately 150 different parameters such as metals, nutrients, cloudiness, and organic and microbiological compounds. These important and various parameters provide an overall judgement of water quality.

6. What tests are done to ensure good water quality?
There are a variety of tests performed to determine water quality, including the monitoring of taste, odour, turbidity (cloudiness), conductivity, dissolved solids, chemicals, microbiological findings like E.coli, cryptosporidium, giardia, and organic findings like pesticides, and hydrocarbons.

7. Who do I contact for more information?
Call 3-1-1 or visit www.calgary.ca for your copy of the 2006 Water Quality Report.

- *Conclusion.* The conclusion may be a statement or a recommendation, depending on the nature of the briefing note. Often, a decision-maker will read only the purpose statement and the conclusion. Be sure that your conclusion is complete and to the point.

Before you write your briefing note, make sure that you have a clear understanding of who will be reading the note and how it will be used, as this will affect its content. Most importantly, the briefing note must be reliable. Every fact must be checked and rechecked, and any dates or numbers should be confirmed. This is not the time to go by memory on key information—be sure to refer back to the original sources to be certain every point in the briefing note is correct. After completing your note, remember that the rules of proofreading apply to every form of professional writing. Do not rely on spell-checking or grammar-checking programs; learn to be a critical editor of your own work.

A briefing note may also be referred to as a policy memo; however, this latter term implies that the writer is considering a wider view of the implications of recommended actions. Additional sections may be required to describe the current policies and the impact of these policies on the recommendations. This can be particularly important for government organizations in the time period following elections or a change in government, as new officials may not be aware of existing policies.

Proposals

There is one reason for writing a proposal—to be successful in obtaining a target. For most proposals, the target is a business contract or project funding. The proposal is a 'pitch' that shows that you have the best plan for responding to the needs of the client or organization, one that is superior to any competitor's bid. You are showing that you can provide a needed good or service, and the recipient is being asked to approve or fund your proposal. A successful proposal clearly addresses a need and offers acceptable options to those who review it.

There are three kinds of proposals:

1. A proposal that responds to a 'Request for Proposals' (RFP) from an organization, with formal and detailed criteria that outline the client's expectations for the bidder. This may be a general advertisement in a newspaper or on the company's website, or one sent only to specific firms with the skills and abilities to bid on the project.
2. A proposal that responds to a less formal request from an organization to provide a bid that addresses a problem or issue. In this instance, the organization may not be sure of what it is looking for from the bidder, but a situation needs to be addressed and funding is available to seek a solution. Generally, the organization will not advertise this but will ask that one or more qualified bidders put together a proposal and indicate how they might address the issue.
3. A proposal that is not requested by the client, but defined by the bidder. The unsolicited 'cold-call' proposal identifies a need for the client that the client may not be aware of, and the bidder provides a solution.

Most proposals fall into the first and second categories. In the first category, the organization defines the issue, scopes out the expected response to the issue, and often provides a dollar figure for how much the organization is willing to spend to address the problem. A formal request for proposals is useful in that it defines the parameters for the bidder, but this can be limiting if the bidder cannot go beyond the scope predefined in the RFP.

For example, the project might call for consultation with the public. The sponsoring organization sees this consultation occurring as a series of open houses, and establishes a budget for these events. However, the bidder sees the merit in developing a website and generating a web-based response. If the idea is included in the response to the RFP and if the budget amount then exceeds other bids, the organization may discount the creative bid in favour of the lowest bid. If you are proposing actions beyond those included in the proposal call, be certain to identify these as 'out of scope' and extra costs, to be included only at the discretion of the sponsoring organization.

In the second category, an organization notes a problem and the bidder is asked to 'put something together' for the organization's review. These more casual requests for proposals have some advantages: often the bidder is not competing against other bidders, and the format allows freedom in designing the project. However, given that little time has been spent on setting out expectations, the bidder must be cautious not to enter into a great deal of work in defining a problem and work program without some assurance of a project (and budget) to complete the work.

The third category, the cold-call proposal, is generally in a written format. The bidder identifies a problem and provides an answer, generally before the recipient is aware that a problem exists. The proposal is mailed to a recipient, and the bidder either waits passively for a response or attempts to follow up the bid with personal contact. At one end of the spectrum, the cold-call proposal could be a simple, one-page statement that sets out the problem and a response. At the other end, the proposal could be highly detailed, containing a great deal of information on the issues and recommended actions. Again, sending out unsolicited bids costs time and money. A great deal of effort may be required to develop the bid without any assurance of available budget or even that your proposal will be seen by the correct individual. Proposals that respond to the first or second categories have a greater likelihood of success.

As with briefing notes, there is no single standard form for proposals—generally, the format differs with each organization. Therefore, when submitting a proposal, it is imperative that the writer have a clear understanding of the expectations of the client or funding organization. If the organization provides any information on expectations in the proposal, read this information carefully, extract every essential item, and make sure that your submission clearly responds to the requirements. Be certain to meet *all* requirements for information and do not exceed the requested length/number of words.

You will save time by developing a standard format for proposals that includes a cover page with your logo and contact information, a table of contents template, and attachments listing your qualifications or past projects. However, each bid must be tailored to the specific requirements of the client and the problem. If a client perceives that you are sending a boilerplate bid, you will not be successful in obtaining the work. Generally, proposals contain the following information:

- *Introduction.* Your introduction should clearly state that the following document is a proposal for a specific purpose. Follow this with a statement that will encourage the recipient to keep reading: show there is a need for the proposal and you have the ability to address the problem.
- *Problem statement.* In a sentence or paragraph, outline the problem or situation. This will illustrate the issue as you understand it. The specification of a problem statement is also an opportunity to verify that you and the recipient agree on the problem.
- *Background.* Why is this a problem? A proposal is not a place to rehash every situation, incident, or event leading to the problem, but an overview will show that there is a need to take action. It may be that the recipient knows full well the background to the issue and will not benefit from a background section. It is recommended that this be included, however, as not all readers may have the same level of knowledge, and this section shows that you, as the bidder, understand the path leading to this proposal.
- *Proposal.* This is the key element of your document. In plain language, spell out what you propose to do for or provide to the recipient. List all 'outcomes' (such as documents you will produce, seminars that will be conducted, and surveys to be completed) and provide an outline in chronological order of how you propose to address the problem.
- *Methods.* Depending on the question addressed in the proposal, the reader may need information on how you propose to answer the question. What forms of qualitative or quantitative analysis will you undertake as part of this proposal. What technology will you use? Use this section to show the reader that you are at the leading edge of research in your field.
- *Benefits of your plan.* By now, your recipient may have some questions. Use this section to debate the merits of your proposal. Highlight the likelihood of success if your plan is followed. Address any negative issues by providing solutions or options.
- *Time frame required to implement the plan.* Provide a schedule with as much detail as possible. Clearly mark points on the schedule where actions will be taken or a product will be delivered to the client. If the project is for a long term and it is difficult to cite specific dates, indicate the time frame expected for each phase. Use this section to show the recipient that you can provide the good or service in the time specified. Do not underestimate how long it will take to complete the project—keep the time frame realistic and work in some 'wiggle room'. Delays are frustrating to the client and will not add to your credibility.
- *Involvement of other individuals/sub-consultants.* If other individuals will be involved in the project, list them now. They may be listed by name, with

resumes included in the section on 'qualifications' (see below) or by job category (geotechnical engineer, social worker). The recipient will then be able to evaluate your team's suitability for the project.

- *Costs.* This section will list your charge out rate (either as a total amount for the project or by the hour), costs for supplies and equipment, and administrative costs such as photocopying and mileage. Hours of work will be shown for each task or phase. A total amount is specified at the end of the section.

  This section of the proposal is often the most difficult to complete. With experience, it becomes easier to estimate the costs and time frames required to complete a task.

  The most common error is to underestimate the budget with the intention of being the lowest bidder in a competitive proposal call. *Do not* underbid the value of your work—a credible bid is one that presents a realistic cost estimate and time frame. Several negative results may be obtained from underbidding: you may be working for a severely reduced rate or at a loss to complete the project; you may be unable to complete the work; or you may provide an inferior result.

  While some texts on proposals suggest that underbidding is a good way to 'get your foot in the door', you will be creating a false perception of the value of your work—it will be difficult to convince a client to pay more for a future project when you were so readily available at a bargain price. When bidding on a project, be certain that you include not only the costs for your time, but also for equipment, travel, office supplies, long-distance charges, specialty work (such as custom photos), and photocopying.

- *The cost of inaction.* Include this section if you can specify specific costs to the reader if immediate action is not taken. This section should include pointed, targeted statements that will create a sense of urgency and may lead to a rapid decision on your proposal.

- *Conclusion.* Reiterate important points and bring the focus back to the problem. Specify why you or your organization is the right choice for completing this project.

- *Qualifications.* Provide a resume, list of previous projects, and references that illustrate experience and information on qualifications (listing projects and outcomes) for you and other project team members. Do not embellish your qualifications: not only is it unethical to lie, but it will be embarrassing and could have long-term career implications if your exaggerations are discovered.

- *Detailed appendices.* If required, include appendices that provide background and technical information that supports your proposal. Attach the information that will allow the decision-makers to support your proposal.

While these are common sections in a proposal, it is critically important to tailor each proposal to the specific issue or format required by the RFP. A proposal must be clearly written and care must be taken to avoid obscurities in defining the problem or proposing solutions.

Practical tip. Vaguely worded promises will not offer credence to your proposal, and may be damaging if the client misinterprets your intention. Strong, specific language is required to illustrate what you can do, how you can do it, how long it will take, and how much it will cost.

A proposal is a document of persuasion—you are writing the proposal to obtain a contract or funding for a project. Your proposal must show that you understand the problem and that you have crafted a way to address it. It must show the benefits of your solution. It must also be cost-effective—for you and for the client.[2] Your proposal has the greatest chance of success if you meet the client's needs. Make it easy for the client to agree to your proposal by proving that you are the best choice for an on-target, on-time, on-cost solution.

Internal Memos

Within organizations, the most frequently used form of written correspondence is the internal memo. A memo provides a written record on the writer's opinion, recommendation, or decision, and is a succinct method of keeping decision-makers, colleagues, and staff informed on actions or issues within an organization. An internal memo is meant to be read only by specified staff within an organization. However, as with all correspondence, the writer should be aware that any printed information could be circulated to a wider audience or the public (see Chapter 5).

As is the case with all professional writing, a memo is written to meet the reader's (generally the decision-maker's) needs and requirements; the writer is providing information needed by the reader in a succinctly worded and fact-based format. While every organization will have a different format for memos, the following outlines a conventional format and the content of each section (note that memo templates are available with word-processing programs or on-line).

Format:	Single-spaced, even margins (5 centimetres), double-spacing between paragraphs, left aligned or justified.
Top of page:	The word 'memorandum' or 'memo' is printed at the top of the page to indicate the format of the correspondence.
Reference Information:	In this first section the reader is notified of the:
	Date: the date the memo is written.
	To: the identified receiver. This may be one person, a list of people, a department, or the entire organization.
	From: the writer. A job title is often included. Note that a memo does not include an opening or closing salutation such as 'Dear Ms. Smith' or 'Yours truly', and the writer does not sign the memo.
	Subject: the topic (instead of the word subject, the abbreviation 'Re:' is sometimes used as the shortened form of 'Regarding'. Make this statement clear—what are you writing about?

Introduction:	The first paragraph, the introduction, provides a brief description of the purpose of the memo. This is a brief, fact-based statement that explains what the writer may expect to read in the memo.
Body:	In the body of the memo the key points are outlined in short sentences or bulleted text. Headers may be used to break the body of the memo into component parts (such as Background, Key Points, Analysis, Results, and Recommendations). As with most professional writing, the text should be written in plain language, with short, powerful sentences that offer no opportunity for misinterpretation.
Final paragraph:	This paragraph may be labelled 'Conclusion' or 'Action' or may flow from the body of the memo. If action is required by the reader, it is specified here.

MEMORANDUM

Date: August 8, 2008
To: All Staff
From: Belinda Apex, CEO
Re: New Memo Review Procedure

Given the recent unfortunate incident where a letter was sent to our primary client, misspelling his name, causing great embarrassment and the loss of the client, the following procedures will be put in place immediately:

- Every outgoing piece of correspondence must be checked for spelling and grammar by the writer.
- The correspondence must then be reviewed and edited by the writer's supervisor.
- Any future errors will be noted in the employee's record.
- All employees are hereby notified that a 'substantial' error (with the level of impact of the above noted error) may be grounds for the dismissal of the employee.

If you have any questions on this matter, contact the Corporate Services Department.

Figure 4.6 An Internal Memo

Note: The use of bold text can assist the reader in navigating the memo by highlighting important text or new sections. Generally, a memo will not exceed one page in length, and like a briefing note the memo may be supplemented by appendices that provide further information.

Practical tip. Don't go overboard in using boldface or italic in a memo. Let your words speak clearly and with power, not your overuse of typographical gimmicks. When you use typographical emphasis, you are effectively trying to tell the reader what's important and what to think, and many readers would like to determine importance for themselves and don't like to be told what to think. Besides, more than sparing use of bold and italic detracts from what really should be emphasized.

While a memo is often viewed as a casual form of correspondence, it remains imperative to check spelling and grammar. In addition, be certain that all names are spelled correctly—it is difficult to use an on-line spelling program to check names, and people can be sensitive about misspellings. The purpose of your memo may be jeopardized if the receiver is offended by a simple mistake.

Business Letters

The most important issue to remember when writing business letters is that you are sending correspondence on your organization's letterhead and the reader sees you as the representative of that organization. Depending on the size and nature of the business, the letter will be placed on file, eventually archived, and will remain available to future researchers in perpetuity. Be certain that the text you produce in the format of a business letter is accurate, defendable, and will stand up to scrutiny.

As with internal memos, most organizations have a format for business letters and templates are available with word-processing programs or on-line for the professional writer. A sample template is shown below.

Format:	Single-spaced, even margins (5 centimetres), double-spacing between paragraphs, left aligned or justified.
Five lines from top of page:	Date (justified to the left) and File Number (justified to the right). The distance from the top of the page and the margins may be adjusted to fit the company's letterhead or to ensure the completed letter is centred on the page. **Leave two spaces.**
Address:	Name and title of recipient (with the correct identifier—such as Mr., Ms., or Dr.). Address of recipient (be certain that this information is correct). **Leave a space.**

Opening salutation:	In most cases, a formal salutation such as 'Dear Ms. Apex' is used. If you are very familiar with the recipient and that person would agree with this familiarity, you may use the individual's first name.
	Addressing the letter to a person is much more effective than addressing it 'To Whom It May Concern' or 'Dear Sir or Madam'. If it's important enough to send the letter, it is worth five minutes of your time to find a person to send it to.
	Leave a space.
Subject or Re:	In some business letters, the subject is identified for the reader. This is true in most government organizations or companies that send and receive a great deal of correspondence. Descriptive information, such as a file number, property address, or legal description, may also be included.
First paragraph:	Be specific. What is the purpose of the letter? In addition, consider the tone of the letter and find a balance between too casual and a soulless form letter. Use plain language and write to be understood.
Middle paragraphs:	Add additional information that explains the purpose of the letter or provides background information.
Final paragraph:	Specify the action you will be taking or the requested action of the reader.
	Leave a space.
Closing salutation:	'Sincerely' and 'Yours Truly' are conventional closing salutations and always appropriate. You may consider developing your own closing salutation, such as 'Peace and Prosperity' or 'With respect for the environment'. However, caution must be used to ensure your created closing salutation is not unintentionally amusing or unprofessional (such as 'Peace Out').
	Leave five spaces.
Signature:	Sign the letter with your usual signature
Author and title:	Immediately below your signature, type your name, either followed by your title or with your title on the next line.
	Leave a space.
Attachments or enclosures:	Indicate if you are including attachments or enclosures by typing 'Attached' or 'Encl'. If you anticipate some confusion with this or want to emphasize the attachments, list them.
cc, c, or bcc:	The final item on your letter is a list of persons the letter has been copied to. The term 'cc' refers to 'carbon copy', a means of duplicating letters with

carbon paper, but is still used as a convention. Alternatively, you may use 'c' which refers to 'copy' and no longer provides the historic reference to carbon paper. Either 'c' or 'cc' notifies the reader as to other recipients of the letter. You may also 'bcc' other recipients, which means to send a 'blind carbon copy'. The primary recipient is then unaware of others who received the letter. The bcc recipients are only listed on *your* copy of the letter.

Cover letters, discussed below, are also a form of business letter. In a cover letter, the contents of the letter are not as important as the attachments. The

Apex Corporation
555 Pyramid Drive, Hierarchy AB T1A 5Y9

August 8, 2008   File No. 5512

Mr. J. Howard
Summit Glass Works
222 Zenith Drive
High Point, AB T1Y 2B2

Re: Proposal for Corporate Art—Main Lobby Installation

Dear Mr. Howard,

Your counterproposal on the above referenced project has been reviewed and we accept your offer.

We look forward to working with you on this important project. Your submission, titled 'A Great Deal of Glass', was the clear favourite among our review panel.

Enclosed is an executed copy of the agreement for your files. We will be contacting you shortly to confirm your schedule and preferred delivery method.

Yours Truly,

Belinda Apex, CEO

Encl: Contract; Corporate Art: Main Lobby

cc: D. Smith, Director of Purchasing

Figure 4.7 Business Letter

cover letter serves as a means of introducing the attached document, acting as a form of a press release. The cover letter should specify the salient points of the attachment(s) (the who, what, where, when, and why), but should provide a summary. Contact information (likely that of the person signing the letter) is also included should the recipient have any comment on the document.

E-mail

The pervasiveness of e-mail is both an advantage and disadvantage for the professional writer. It is cost-effective, efficient, allows for rapid responses, and many people can be notified simultaneously. It is easy to respond to a colleague or client without a great deal of thought—and therein lies the problem.

In a professional setting, e-mail should be given the same level of attention and critical thought as any written correspondence. Correct spelling and grammar are just as important. The use of informal abbreviations or Internet icons (such as LOL, happy faces, or emoticons made by combining text characters) should generally be avoided—what is appropriate in non-professional e-mail or web-based correspondence is not always acceptable in a business situation. Casual language and a 'chatty' style of correspondence are also not recommended. In addition, be aware that written correspondence can be misconstrued by the reader—the use of sarcasm or a pun may be misunderstood, causing confusion or embarrassment, and these forms of writing sometimes do not translate well. Be sure to write clearly and proofread to look for places where the e-mail could be misunderstood.

Some organizations may take a more informal approach to e-mail, but it should be recognized that once an e-mail is sent, the writer no longer controls its distribution or printing. Care must be taken in both style and content.

To ensure that an e-mail meets professional standards, consider the following:

- Write with the same care and attention as you would with any business correspondence. Do not let the format cause you to use more casual language than would normally be expected in business correspondence.
- If you are using e-mail templates to add colour to your e-mails, verify that the e-mail will be received in the correct format by sending it to yourself first. Confirm that the e-mail views as intended on the screen, and that the template does not make printing difficult.
- Use the subject line to your advantage—instead of listing 'new policy' as the subject, let the reader know that the e-mail is about the 'New policy on staff evaluations'.
- Do not use unusual fonts that may not be available on the recipient's computer.
- Do not use unnecessary graphics (graphics can consume too much memory on some devices).
- Do not send attachments unless you are certain of their origin and can confirm they are virus-free.
- Ensure that links to websites are operational by testing them before sending the e-mail.

On occasion, you may receive an e-mail that causes you to be angry or upset. Before sending back a 'howler' of your own, carefully review the e-mail to be certain that you understand it correctly. Sometimes, the use of sarcasm can be misunderstood or the meaning of a pun can get lost between the reader and the writer. Decide if an e-mail response is the best choice, or if a telephone follow-up would be better.

If you draft an e-mail or a response that could be construed as controversial or is written when you are in a highly emotional state, save it in your 'drafts' folder for 24 hours, then review it, or ask a trusted colleague for an opinion. If the e-mail still reflects your professional opinion or position on the subject and you are prepared to defend it, then send it. Just be certain that you have not let the speed and ease of e-mail cause you to send correspondence that you will later regret.

Resumes

While not strictly a format of professional writing, a resume may be the most important written document that you create (and the one that enables you to get the career where you do professional writing!). A resume is a descriptive document that best represents you to potential employers—consider it to be an advertisement of a sort, but one that must reflect you and your background in the best possible manner.

A resume outlines your career experience, skills, and education in a way that will intrigue a potential employer and cause him or her to contact you for an interview. It can be used as a predictive tool, illustrating how well you will do in a potential career position. Ultimately, a resume is the document that gets you in the door to a desired career or job. Without a good resume, landing in your desired career is much more difficult.

Not so long ago, the rules for resumes were extremely strict (not more than two pages, established format listing previous employment and education only, three references). Today, given increased diversity in the workplace and the varied experience of workers, resumes may take many formats. On occasion, a video resume may be requested of you for certain jobs, but most often a paper resume remains the industry standard. If the employer requests an e-mailed resume, send a PDF version or other format that allows the formatting of the resume to remain intact, regardless of the program it is viewed in. If the employer requests a word-processed version, send both types to ensure that the resume, when printed, looks as you expect it to.

The following lists items that should be included in a resume, items that should never be included, and suggestions on how to make a resume stand out from countless others received by perspective employers.

Contact Information

It should go without saying, but a resume must contain accurate and up-to-date contact information. Be certain that your current address is listed on the resume, and be aware that a potential employer may infer information from your address.

For example, if you are applying for an out-of-town position and the employer has received similar quality resumes from in-town applicants, the employer may favour the local employees over those more distant (thereby saving moving costs for relocating the employee).

Your telephone number must also be correct—be certain that the phone number given is one that you regularly monitor (use your cellphone number over a land line if you are much more likely to be reached on your cell). Also, your voice-mail message might be hilarious to you, or to your friends and family, but it may not be so amusing to a potential employer. An employer is most likely to contact you by telephone for an interview, so professionalism is critically important.

As for e-mail, again, be certain that you provide an address you regularly monitor, and also consider if your e-mail address is descriptive of your current situation. It is likely that the address created in junior high is no longer representative of a university student. In addition, be cautious of e-mail addresses that are leading or overly descriptive: an e-mail that reads 'canuckswincup2010@hockeyismylife.com' may not be well received by a rabid San Jose Sharks fan or an employer who might be concerned with your level of commitment to the workplace.

Facebook and other information-sharing sites, as well as personal websites, have created new issues for job seekers; the more available you are to the world, the more available your site is to a potential employer. Consider if the information contained on the site accurately represents your current status and if it will present well to a potential employer.

Career Objective

Immediately following your contact information, include a job objective. This is a one sentence targeted statement that is recreated for every new position to which you apply. Do not have a generalized statement such as 'Interested in an interesting career in an interesting organization', which is meaningless, obviously generic, and shows that you don't write well. Instead, use the job objective as a means of grabbing the employer's attention and focusing attention on the resume that follows.

Career Experience

Career experience may be listed chronologically or grouped around topic areas (for example, skills in facilitation followed by a listing of projects that illustrate this skill); however, most resumes use the chronological format.

Work experience is listed in order, with most recent experience first, then working backward through previous positions. Generally, the job title, employer, and time spent at that organization are listed, along with key duties and highlights of career experience. For example, you may list general tasks (report writing, interaction with clients, product sales of $1.2 million in the last fiscal year) as well as emphasizing any awards or recognitions received (e.g., salesperson of the month for four consecutive months; awarded the Toeffler Prize for Innovation).

The objective is to present enough information to show that you should be considered for the applied-for job.

It may be that you have very little career experience or have gaps in your experience. Summers spent travelling and school years focused on maintaining a high grade point average may have left little time for employment. If this is the case, *do not* consider embellishing your resume to create a better or fuller work history. Several key questions at an interview and a reference check will likely reveal the fabrications; even if you manage to be hired, false pretenses are grounds for dismissal. Instead, fully develop the career experience you do have, and be certain to include any volunteer work or even academic experiences (like major projects or field schools) that will help to give a full impression of you to the potential employer. For example, if you have participated on a major project and the results were published in a peer-reviewed academic journal, note your involvement (without over-inflating your role). This will assist in highlighting both your ability to contribute to a large project and to contribute successfully to an initiative that required substantial teamwork. A field school experience, where you participated in a research project or volunteered in a community, should also be noted in your resume.

Volunteer experience qualifies as career experience. Be certain to identify it as volunteer work so it does not look as though you are padding your resume, but include it. Volunteer work shows commitment to a cause you are interested in and also that you are willing to contribute back to society. Once your resume is sufficiently developed with a broad range of career experiences, volunteer work can be separated out into its own section.

Education

Again, a chronological format is typically used for this section of the resume, starting with the most recent education first. If you are currently enrolled in a degree program, list it along with an anticipated date of completion, such as:

Bachelor of Arts, Honours History (*anticipated graduation Spring 2010*)
Vancouver Island University

A completed degree would be shown as:

Bachelor of Arts, Honours History (*2010*)
Vancouver Island University

List any awards or scholarships, detailing the dollar amount. If it works to your favour, list your current or final grade point average or class standing. As well, if you graduated with honours or another special citation, be certain to note it.

Education is normally only listed back to high school, and after a few years of career experience the reference to your high school may be excluded. An employer is likely much more interested in what you have done after graduation—list all training, including courses taken in the workplace or other classes you may have taken outside of the regular curriculum at university.

Generally, if your resume does not contain a great deal of career experience, the section on Education will follow your Career Objective.

Life Skills and Achievements

This section details those skills that may not be related to a specific previous job, but should be highlighted. Computer skills should be included—list the programs you are familiar with that would be relevant to the position you are applying for. Any certifications should also be noted, as the completion of a program suggests that you have the ability to see a project through to its conclusion. Awards and scholarships may also be highlighted, if they have not been included in the Education section. In addition, languages spoken may be highlighted (this can be increasingly important in a compressing world).

It is not relevant, however, to list awards gained previous to graduation from high school (such as a Grade 9 scholarship or an elementary school poetry award). From a chronological perspective, an employer may be interested in scholarships or awards obtained in the transition to university or gained at the university level. Unless highly relevant (a lifesaving award granted in junior high school listed as part of your application to be a police officer), earlier experiences tend not to add useful information to a resume.

Memberships

If you are a student member of a professional organization, list it here. If you are not yet a member of a professional association, consider registering in an organization in your area of interest. The cost for student membership is generally reasonable and membership will give you access to the 'members only' area of the organization's website as well as to publications, job postings, and conference information. These organizations often have mentorship programs as well, and may be able to connect you to persons in your area of interest, potentially leading to job opportunities or lifelong connections.

Other affiliations may include any range of interest groups: while a resume must be truthful, you may not want to list all social affiliations if they may not be positively viewed by a potential employer. Consider if discretion may be more apt than complete transparency.

References

For each job application, consider if references need to be listed. If they are requested by the employer, then provide references. However, it may be that you are currently in a position and would prefer that your employer not know that you are seeking other employment. In this situation, if references have not been specifically required, then state 'references available on request' at the end of the resume or within the cover letter. You can then explain your situation at the interview, and ask that your references only be contacted as a final step in the interview process.

When references are included, often three is the requested number (although employers sometimes ask for a range of references—one academic, one personal, and one professional). Clearly specify why the individual is being used as a reference (for example, 'former employer' or 'academic supervisor') and provide all necessary (and up-to-date) contact information. Do not use relatives as references, unless you have been employed at a family-owned business and no other references are available. A reference with the same last name is generally viewed with some suspicion.

Also, be certain to ask your potential references if you may use them as references, and periodically reconfirm their willingness to stand as references. It serves no useful purpose if a reference is either surprised by a reference check and poorly answers the questions, or worse, if the reference refuses to act as a reference when contacted by a potential employer.

Be Accurate, Be Current

Finally, check for errors in grammar and spelling. Print a copy of your resume and read it word for word and sentence by sentence. Ask a colleague (perhaps more than one) to proofread your resume. Some employers will immediately reject a resume containing any errors, regardless of the skills or experience of the individual. A resume must be completely free of error in spelling, grammar, and sentence structure.

While it may seem onerous, mark a day once each month to review your resume and make any necessary amendments. Most people do not consider the state of their resume until they need to apply for a job, and then the task of remembering past successes and the specifics of past employment can be difficult. A small amount of time spent each month will keep your resume up-to-date and reflective of your skills and accomplishments.

The following provides a template for the development of a resume (there are many good examples of resumes tailored to specific careers available on-line at job search websites—to ensure your resume does not appear dated, periodically review on-line resumes to see if there are new formats or sections that would work for your resume):

Resume Tips

1. *Use words that show action.* Words like 'create', 'develop', and 'implement' are powerful and add life to your resume.
2. *Use numbers.* Where relevant, numbers can add a quantifiable measure of success to your resume. If you have positive sales figures or exceeded timing targets in a previous job and those numbers would be relevant to the desired job, add them to your resume.

Name

Address 1
Address 2
Telephone (most likely number)
E-mail (most frequently checked)

CAREER OBJECTIVE

A targeted statement to attract the attention of the employer (this is exactly who we are looking for!)

EDUCATION

- List degrees—including degrees not yet completed (add an anticipated completion date)

CAREER EXPERIENCE

- List previous employment, starting with the most recent position first
- State the job title, organization, and time spent at the organization
- List key tasks and achievements, using numbers where relevant
- Use action words and short statements, highlighting your strengths

SKILLS AND ACHIEVEMENTS

- List of accomplishments
- Awards
- Languages spoken
- Relevant computer programs (for example, skills in GIS or statistical packages)

REFERENCES

- List three references or state 'available on request'

Note the use of white space—the resume is not overcrowded with text.

Be sure to include training programs or certificates (either here or under 'skills').

If there are gaps in your resume, be prepared to discuss what you were doing at that time (attending school, travelling).

This section is a good opportunity to highlight why you would be the right person for the job.

Be sure your references are aware of the job application and be sure that you have a general sense of what they will be saying about you.

Figure 4.8 Resume

3. *Do your research.* Put some effort into researching your potential employer, not only for the job interview but also before you tailor your resume to the position. Check the company's website for recent press releases or annual reports that provide information on the current state of the employer. Review newspapers to see if the employer has been faced with any recent issues or controversies. For example, if the employer has been awarded 'Business of the Year', you may want to mention that in your cover letter as one of the reasons why you are interested in potential employment opportunities. Any research completed will also be useful as you proceed into an interview.

4. *Printing.* Print your resume on good-quality paper. While copy paper is minimally acceptable, a better quality of white or slightly off-white paper will make your resume feel more substantial and will separate it from the others being reviewed. It is not recommended that brightly coloured paper be used for a resume, or that coloured ink be used to highlight different elements or sections. Industry standard in North America is black ink on a white or cream-coloured paper stock: it is unlikely that you will ever be penalized for complying with this standard.

5. *Hand delivery.* If at all possible (unless you are applying for an international position), hand-deliver your resume to the person to whom you will be reporting. By meeting the person in advance and making a good impression, you are increasing your chances of getting the job—the interviewer will have met you prior to the interview, will have a positive impression of your social skills and 'fit' with the organization, and will be much more likely to support you for the job. Research the company website or call the reception desk to find out who you need to talk to.

Box 4.2 What Not to Include in a Resume

- **Do not use a character font.** A resume is not the document to prove your individuality and why you do not need to conform to anyone's standards. A character font can be difficult to read, and does not illustrate professionalism.
- **Do not include a section titled 'hobbies and interests'.** The section on memberships provides information on your professional affiliations. Unless you are applying for employment at a craft store, employers are unlikely to be interested in your skills in knitting and decoupage.
- **Do not include personal information.** In Canada, an employer does not have the right to personal information, such as birth date or marital status, as part of the interview process, and this information should not be used in the review of resumes.
- **Generalities.** When you can, be specific. Stating: 'Generated $1.2 million in product sales, exceeded company sales targets by 15%, and was awarded "Employee of the Year"' is better than saying 'Worked in sales'.

Cover Letters

A cover letter introduces your resume to a potential employer and offers an additional opportunity to make a good impression. The cover letter should clearly specify your contact information and the position applied for, along with the application number (if one is listed in the job advertisement).

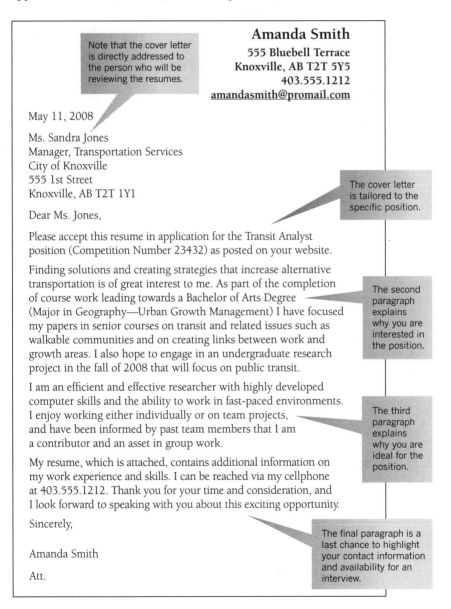

Figure 4.9 Cover Letter

The cover letter must be written specifically for each job application. A generic cover letter is easily identified and will be viewed negatively by an employer faced with the task of reviewing a stack of resumes (particularly for jobs posted on-line, the number of resumes received may easily be in the hundreds). The font, style, and paper should be the same as your resume.

Do not address the cover letter 'To Whom It May Concern' or 'Dear Sir or Madam' unless you do not want your resume to be read. Find out the name of the person who will be at the interview or will be supervising the position and address the resume to that person. At the very least, address the resume to the person in Human Resources who will be receiving the resume.

The cover letter should be considered an introduction into the organization and an additional opportunity to prove that you are the ideal candidate for the job.

Chapter Review

This chapter reviewed formats of professional writing that will likely be encountered in the workplace, including research reports, proposals, newsletters, press releases, and briefing notes, along with recommendations on how to achieve precision in professional writing. As with academic writing, every organization will have established criteria for different formats. Be certain to meet the formatting requirements to ensure that your work is evaluated on its content, not its structure.

Review Questions and Activities

1. Review (or complete) a resume following the format outlined in this section. In your calendar, set up a reoccurring appointment once each month, blocking out a half-hour to review your resume and update it with current activities.

2. Research professional organizations in your field of study. Most organizations provide low-cost memberships to students. If possible, become a member to gain access to the organization's conferences, newsletter, and potential job opportunities.

3. Review a newspaper and seek out articles that are clearly based on press releases. The articles will follow the inverted pyramid format discussed in this chapter, as well as other criteria (e.g., a quote in the third paragraph, an announcement of an event).

4. Write a press release announcing an event for an organization you are involved in (with the organization's permission; note that many large organizations have Communications Departments that are responsible for press releases) and submit it to a relevant newspaper. Ultimate success is achieved if your press release is published 'as is' by the newspaper.

5. Obtain a sample of newsletters (both paper format and on-line). What elements are appealing? What does not work? Create a template for a newsletter that uses the most attractive elements from the reviewed documents.

Chapter Five

More on Writing

Sources and Citations

In both academic and professional writing, the documentation of sources proves that a written work has scholarly and intellectual merit. A reference list with depth and breadth indicates that the writer has thoroughly researched the topic, understands the background to the issue, and can speak with some knowledge on the subject. While there is certainly a place in both styles of writing for opinion pieces that provide only the author's point of view, a well-referenced document illustrates that the author has considered more than his or her own ideas and has developed the writing around the scholarship of others. Evidence from previous research is compiled, the evolution of an idea is traced, other views are documented; the result is added intellectual weight and legitimacy.

A reference list is developed from the sources either read as background to a document or directly quoted within the document. Quotations (or citations) are used to reinforce an argument, establish the context for following comments, or summarize a key point. They add a sense of authority to writing, particularly if they are from sources known and respected by the reader. When quotations are used, they must be correctly referenced and documented to ensure that the original source is credited with the text.

Properly done citations offer the following benefits:

- The knowledgeable reader can evaluate if your document is current and complete.
- In an academic document, the reader can site your purpose or thesis within your scholarly field.
- The dangers of plagiarism are avoided as sources are documented.
- Both the reader and the writer can reference back to source material.
- Citations 'confer authority'[1] to the text—ample citations imply that the writing is well researched.

In brief, academic and professional writing will only become better with the use of citations.

As a student, most of the sources investigated in the course of your research are published materials—books, journals, and articles. These sources can be easily documented and referenced in your written work. However, as a professional, you may encounter a much wider range of information and ideas that will influence your work. As noted in Box 5.1, myriad sources may shape your ideas—contents on a web page, a casual conversation, or the review of an unpublished manuscript. While these are not published sources that have undergone academic or professional scrutiny, they still must be cited correctly.

Box 5.1 Sources of Information

Books: Written by one or more authors, or a collection published by one or several editors. Your citation references the author of the work quoted as well as the source.

Journals: Academic journals (either in printed form or on-line)—both peer-reviewed and non-peer-reviewed.

Statistics and numbers: Generally, quoted statistics are from a government agency or a reputable source. Quoted numbers must also be from a legitimate agency. Be cautious of using data that cannot be verified or triangulated with a second or third source.

Government reports: Much of the information published by governments is not credited to any writer. The citation then references the government agency and the document.

Think-tank reports: Many non-governmental institutions, which are funded by donations, grants, research agreements with mainstream media outlets, and, in some cases, their own publications, have research staff and writers and/or fund 'independent' research, and publish reports and monographs on issues of public concern. You need to be aware of the ideological bias of the institution, just as, in some instances, you should know the bias of individual academic writers.

Minutes from meetings and conferences: The proceedings from meetings or conferences are often made available after the event. If you were not present at the event, verify whether the written proceedings are verbatim (what was actually said) or a summary (the writer's interpretation of what was said). Follow up with the writer or speaker if accurate information is needed.

Magazine and newspaper articles: The use of non-academic sources must be carefully considered for validity. A series of well-researched feature articles on an important social issue is rather different from letters to the editor or articles on the op-ed page.

Audiovisual sources: While not in a published format, most audiovisual presentations are subject to copyright laws, and the use of ideas or quotes from these sources must be documented.

(Continued . . .)

Lectures and presentations: If your ideas were obtained from your attendance at a lecture or presentation, the source of the information must be cited. If no written documentation is provided of the lecture or presentation, note that your citation will reference your interpretation of the idea. If possible, confirmation of the speaker's intent should be obtained (as well as permission to use the source in your written work).

Discussions/conversations: On occasion an idea may come from a casual conversation or a discussion with learned colleagues. If the idea cited is not your own, the use of the idea in your work should be discussed with the originator. Permission to use the idea must be obtained.

Unpublished works: If your idea is obtained from a draft document, unpublished dissertation, or unfinished paper, it must still be documented. If possible, obtain written documentation that specifies that you have the author's permission to use the information.

Internet: With e-mail or information obtained from a web page, remember that web pages can be edited over time, so document the date of retrieval. E-mails, as well, can be edited or deleted. Printing the e-mail (noting the date of writing as well as the date of printing) will provide legitimacy to the citation. Permission of the author may again be required.

In some cases, the reader and the writer may never be able to trace back to the original source to verify the information (for example, if you are quoting a speaker from a conference and conference proceedings were never produced). However, the point of the citation is to show that the idea originated with another individual and to give recognition to that person for his or her work. The use of citations will add to the credibility of your work and provide acknowledgement to the intellectual property belonging to the source.

Many formats can be used for citations. It is important to understand the format used in your discipline or by your instructor. While the differences may seem minute between different formats, it makes little sense to lose marks or credibility because a reference is incorrectly cited. Regardless of the format, there are commonalities in the information required for a citation:

* *Author(s)*. Who wrote the material? If it is more than one author, be certain that all are credited. On occasion, the second and subsequent authors can be lost in a citation. In the physical and medical sciences especially, sometimes there will be many authors listed, as if everyone wearing a lab coat that day was included for authorship. APA (American Psychological Association) style, which is often used in the social sciences, calls for listing the first six authors named, followed by 'et al.' (Latin: *et alia*, meaning 'and others'). For the sake of economy, other citation styles, when there are more than two authors, often list only the first (or senior) author, followed by 'et al.'. Also, be certain that you are crediting the author of the work and not the editor. Edited texts

sometimes are a compendium of written works by a collection of authors. The editor may also be a writer in the text, but the citation must correctly reference the author. Government publications often do not reference an author. Credit the department or agency in the citation. Websites, as well, may list the graphic designer of the site but not the writer of the text. In this instance, the source organization is cited as the author.

- *Year of publication.* While this would seem simple, the publication date of material may take some research. Are you citing the original publication date, a reprint edition, or a later (and therefore revised) edition? For a web page, are you citing the date of your retrieval of the information or the date of publication on the page? As a rule, the date cited is the date of publication— when did the information become widely available to the public? Where no date is listed on the copyright page of a book and you can't ascertain the date, use the abbreviation 'n.d.' (no date) in your reference citation. If it isn't listed but the date is known, you include it in square brackets: [1974].

- *Title.* Cite the entire title of the work. When a title is subtitled after a colon, cite the subtitle as well. If, for graphic design reasons, a colon has not been used on the title page between what are obviously the title and subtitle, in citing the work you should include the colon. Thus, in citing this text, for example, the title would be: *A Field Guide to Communication: For Students in the Humanities and Social Sciences.*

- *Publisher and place of publication.* Not all cited works will have been published, so it can be difficult to include this information. If a publisher cannot be identified, document the source of the information; if a place is not specified, reference the place where the event occurred (that is, for a conference, reference the location of the event). The abbreviation 'n.p.' is used when no publisher can be determined.

- *Edition.* As noted above, it is important to indicate the edition of a published text, as the contents can vary widely among editions. On occasion, new authors are added to later editions, and publication information may change. A well-documented citation will assist in tracing back to the correct edition.

- *Volume/issue number/page number.* Published journals reference a volume and/ or the number of the edition. Again, this information is cited to allow the reader to easily refer back to your source. The page numbers for the beginning and end of the article are specified, unless the citation is to a specific quotation you have used, in which case you include only the page number(s) the quoted material came from.

- *Format.* If the citation references something other than a book or journal, the type of work may be identified if this will be informative for the reader. For example, if the citation refers to a video documentary, a conversation, or an unpublished dissertation, this reference will provide useful information to the reader.

As you do your research, be certain to collect all necessary information to complete your citations. If a source is not correctly cited, you may not be able to trace back to the source or retrieve related information.

As mentioned above, the format often used in the social sciences is a style defined by the American Psychological Association. APA style was developed to establish consistency in the way references are cited, and to ensure that non-traditional references are adequately credited. APA style applies to more than just the citation of sources. In addition, standards for the format of documents, punctuation, headings, and tables and figures are also available. The Association publishes a manual on this format that is likely available in your university library (or for purchase through the APA website). However, for this section, only the specifics around citing references are examined.

Formats are established for virtually every type of citation, including websites. For example:

Books are cited with the author first (last name then initials), followed by the year of publication, title of the text, place of publication, and publisher, as shown:

> Serote, E.M. (2004). *Property, patrimony and territory: Foundations of land use Planning in the Philippines.* Quezon City: University of the Philippines School of Urban and Regional Planning.

Journal articles are cited:

> Ver Eecke, W. (1999). Public goods: An ideal concept. *Journal of Socio-Economics* 28 (2), 139–152.

A **chapter from a text** is sourced as:

> Wolman, H. (1995). Local government institutions and democratic governance. In Judge, D., Stoker, G., and Wolman, H. (Eds.) *Theories of urban politics* (pp. 135–159). Thousand Oaks, Calif.: Sage Publications.

And a quote from a **website** (which appears to have no author or date of publication) is referenced with the first few words of the quotation and/or the paragraph number (number of paragraphs counted down the web page). It is important to reference the date that the item was retrieved from the websites, as websites change and the reference may be lost:

> Specifically, the Nanaimo Estuary covers a wide marine area (n.d.) Retrieved May 27, 2008, from http://www.nanaimoestuary.org.

Note: 'n.d.' means that 'no date' is shown on the website.

Other citation styles include Modern Language Association (MLA) style, Chicago or Turabian style, and American Political Science Association (APSA). Before researching a paper, be certain of the expected format for citing references and adopt a 'no errors' approach to recording references. Always, with citations, references, and bibliographies, the primary concern is consistency.

Computer programs are available that allow the user to enter in source information, and then select a format (often a university will adopt a format and make it available to students and instructors—End Note and RefWorks are two programs used across North America). The information is then reorganized to meet the selected format. As a professional, it is important to research the citation style required by your organization or client. Documenting all relevant citation information as you access each source and the use of citation programs will enable you to switch among citation styles as required.

Practical tip. Some confusion often exists between the use of the terms 'References' or 'Reference List' and 'Bibliography'. The terms are not interchangeable. A list of references includes all the references actually cited in a document, listed in alphabetical order (not in order of appearance). A bibliography refers to all the texts, articles, and other sources read by the writer in the preparation of the document. The source may not be cited within the document, but the bibliography provides evidence that the writer has read a breadth of work in the subject area.

Plagiarism

A common thread runs through the definitions of plagiarism from North American academic institutions: *plagiarism is copying another's work and presenting it as if it were original work created by the author/student.* Many colleges and universities have policies that define what constitutes plagiarism at that institution (Box 5.2). Students are advised to be aware of the definition and policy, and of the penalties that will be incurred if plagiarism is detected in the student's work.

At a post-secondary or professional level, the reader assumes that any portion of a work not attributed to a source is the writer's own. Therefore, as a student or professional, it is unethical to allow the reader to falsely infer that your work is original when it is not.

Plagiarism is the theft of intellectual property and a moral crime against originality. While it is not specifically defined under Canadian criminal law as theft, it remains illegal when found to be in conflict with copyright laws or can constitute fraud when the plagiarism is found to be intentional.

There are four kinds of plagiarism:

1. *citing* someone else's work without reference;
2. *collaborating* when the assignment is to be completed entirely by the student;
3. *self-plagiarism*, where a student uses his or her own work for more than one course, contrary to the regulations or policies of the university; and
4. *purchasing work* and using it for a mark or certification.

Each of these kinds of plagiarism is described below.

Box 5.2 Definitions of Plagiarism from Selected Canadian Universities

Vancouver Island University
'Plagiarism is the intentional unacknowledged use of someone else's words, ideas, or data. When a student submits work for credit that includes the words, ideas, or data of others, the source of that information must be acknowledged through complete, accurate, and specific references, in a style appropriate to the area of study, and, if verbatim statements are included, through quotation marks or block format as well. By placing their names on work submitted for credit, students certify the originality of all work not otherwise identified by appropriate acknowledgments.' (*Student Academic Code of Conduct*)

University of Alberta
'No Student shall submit the words, ideas, images or data of another person as the Student's own in any academic writing, essay, thesis, project, assignment, presentation or poster in a course or program of study.' (*Code of Student Behaviour*)

Ryerson University
'Claiming the words, ideas, artistry, drawings, images or data of another person as if they were your own. This includes:

- copying another person's work (including information found on the Internet and unpublished materials) without appropriate referencing;
- presenting someone else's work, opinions or theories as if they are your own;
- working collaboratively on an assignment, and then submitting it as if it was created solely by you.' (*Student Code of Academic Conduct*)

Dalhousie University
'. . . plagiarism (is) the submission or presentation of the work of another as if it were one's own.' (*Policy on Academic Integrity*)

The policy cites examples of plagiarism including:

- failure to attribute authorship when using a broad spectrum of sources such as written or oral work, computer codes/programs, artistic or architectural works, scientific projects, performances, web page designs, graphical representations, diagrams, videos, and images;
- downloading all or part of the work of another from the Internet and submitting it as one's own; and
- the use of a paper prepared by any person other than the individual claiming to be the author.

Uncredited Use of the Work of Others

Technology may be contributing to the problem of plagiarism. Prior to 1990 most researchers (students and professionals) did not have access to the Internet, and even if they had, they wouldn't have found a great deal of interest. Research was conducted in libraries from available texts. Knowledge was limited to what was available at that institution: the cost of obtaining printed materials would limit the number of academic journals available on site to what the institution could afford, and the purchase of new texts was also controlled by budgets. While inter-library loans were possible, a student was largely limited to the books and journals available at that university. Universities with a strong research focus would pride themselves on the extent of their library collections: from rare texts to current obscure journals, universities and specific departments and faculties considered their library collections to be an important component in attracting the best and brightest students.

Today, any individual with a computer and Internet connection can access an ever-expanding world of information. Information is no longer geographically based, and even rare materials (such as historic documents) can be scanned and made available to academics worldwide. While much of the content on the web cannot be considered academically valid, the Internet does allow virtually unlimited access to on-line academic journals, text excerpts, original scientific research, and experts' websites. (As to the accuracy of these sources, both students and professionals should confirm and know something about the status and reputation of the originator of the data and the rigour of the information standards. A correctly referenced citation that sources incorrect information is not plagiarism, but it is still wrong.)

These are useful and important sources for the academic and professional writer. However, the ease of 'cutting and pasting' may be contributing to increasing intentional plagiarism as the writer 'borrows' substantial portions of a text without citation.

Paraphrasing also is an issue. To *paraphrase* means to restate text in different words, generally to add meaning or to clarify the content of the text. That is, the paraphrasing provides additional explanation to help the reader to understand the concept or content of the paraphrased text. As an example, see the previous two sentences: the last sentence paraphrased the one before it—it offered no new information, just explained the former sentence to clarify its meaning. When paraphrasing is done correctly, the student uses his or her own words to discuss a concept, and the original text is correctly cited. When paraphrasing is done incorrectly, the student will move around only a few words or reorder sentences, but the text remains largely the same. The problem is exacerbated if the original text is not referenced in the student's work.

Miguel Roig, a psychology professor at St John's University in New York City, published an article titled 'When college students' attempts at paraphrasing become instances of potential plagiarism'.[2] In this article, Roig notes that many students have difficulties in identifying plagiarized text versus paraphrased text.

This difficulty is compounded if the original text is highly technical or difficult to read (p. 979). His website (facpub.stjohns.edu/~roigm) provides a guide titled 'Avoiding plagiarism, self-plagiarism, and other questionable writing practices: A guide to ethical writing', which is an excellent source for further information on the difference between plagiarism and paraphrasing.

Unintentional plagiarism may be on the rise, as well, where the author, after reviewing a vast amount of written information, loses sight of the origin of an idea and presents it as original work. The term 'cryptomnesia', coined by the psychologist Carl Jung, refers to a state when an author writes something he or she believes to be original, but is actually recalling a work previously read. Critics of this idea of forgotten memories suggest that it provides a convenient excuse for a plagiarist: the author cannot be guilty if she does not know she is plagiarizing the work of another. It can be difficult to detect cryptomnesia in one's own writing: if possible, have your work read by a colleague with a high level of knowledge of the subject area; this person may best be able to detect overt similarities in ideas or passages of writing.

Some universities are adopting safeguards against plagiarism with software that can detect similarities among a student's work and information available on-line. The ease of 'cutting and pasting' becomes a two-edged sword: as easy as text is to copy, it is just as easy to find the original source document. If detected, the penalties for plagiarism can range from a requirement to redo the assignment to full expulsion from the institution.

To guard against plagiarism, adopt two simple rules in your writing:

1. Any time more than two consecutive words are obtained from a source, that source must be cited.
2. If a source provides an idea that shapes your work, that source must be cited.

If you are uncertain whether or not your work constitutes plagiarism, review your college or university's policy on plagiarism and consider it against your text. If you are still unsure, err on the side of caution and give credit where it is due.

Box 5.3 Plagiarism Software

Detection software is available for purchase on-line for individuals, or may have already been adopted by your university. Check the student calendar or with your instructor to see if it is in use at your institution. A few of the more common packages are described at:

- www.plagiarism.com
- www.canexus.com
- www.copycatchgold.com
- www.turnitin.com
- www.mydropbox.com

Sharing Information

A 2008 investigation into academic misconduct at Ryerson University brought the issue of sharing information into new focus. A group of students in a chemistry course had created a Facebook site for collaborating on assignments and general discussion on the course. The management of the site was taken on by one student, and a group of students used the site to exchange ideas on completing a difficult assignment. The site was discovered by the university, and the student managing the site was brought up on 147 counts of academic misconduct: one for managing the site and 146 for each visit to the site by students on that project.

Ryerson's Policy on Intellectual Honesty includes the statement: 'In the absence of specific approval from the instructor of a class, all students should assume that all assignments are to be completed independently, without any form of collaboration'; therefore, the university considered the managing student's activities to be a form of cheating. It was this clause in the Student's Code that was at issue. The students had been directed to complete the assignment on their own; clearly they had not.

While students across Canada argued the Facebook site differed little from a conversation at a cafeteria table or a discussion in the library, the fact remained that the discussion and the participants were digitally recorded on the site. The university eventually dropped the charges, but the issue highlights this subtle form of plagiarism. If the instructor directs that each student complete an assignment on his or her own, any form of collaboration is fraud.

In a professional environment, collaborative efforts are typical of most workplaces. The difference is that the professionals have not been instructed to work alone and independently arrive at a solution to the issue, and are likely not receiving a mark that will assist them in acquiring academic qualifications. However, the intellectual theft of ideas for credit (and ultimately greater pay or prestige) can be rampant in some workplaces. Being conscious of and maintaining high ethical standards is important in the workplace and the best means of addressing issues of plagiarism.

Self-plagiarism

Many universities also have policies on 'self-plagiarism', or using one's own work for more than one course (and for multiple credits). If an instructor suspects self-plagiarism, it can be tricky to prove that fraud has taken place. It may be that a student is drawn early on in her or his academic career to a particular topic area and writes a series of papers over several terms that build on previous research (yet each paper is distinctly different and a clear progression is shown). Instead of 'standing on the shoulders of giants', the student is standing on her or his own shoulders (if that's possible!) to advance a research program. In this instance, it would be difficult to consider the series of papers to be self-plagiarism. However, if a student writes a paper in one course and then resubmits the same paper in another course with largely the same content and limited additional research, self-plagiarism could be the charge.

As with the plagiarism of another's work, honesty is the key. The level of effort expected for an assignment can be surmised from the student's academic year, the proportional mark of the assignment, and by discussing the assignment with the instructor. If a student is unsure of whether or not the use of a previous paper constitutes self-plagiarism, the issue should be brought to the attention of the instructor.

As a professional, reports and studies will often build on previous work, and self-plagiarism tends not to be an issue. However, care must be taken if information is being prepared for more than one client and presented as original work for each client. In many industries, there may be close connections between clients, and they will not be pleased to know that each paid for exactly the same work.

Purchasing Work

One final area of plagiarism involves the purchase of work completed by another and passed off as one's own. There are thousands of websites that offer papers for purchase: some will provide an already completed research paper on virtually any conceivable topic; some will offer a paper in exchange for another research paper; others will conduct original research and complete a thesis or dissertation for a cost per page. Many of these sites provide disclaimers stating that 'these papers are to be used only as examples' or 'to be used for educational purposes only' (whatever that means). While the cost of purchase may be fairly high, the desperate student may feel driven to buy his or her way through a course.

This is not a new phenomenon. In pre-Internet days, the trade in papers in the frantic final weeks of a term could be brisk, and students have long generated income by doing the work of others for pay. However, a historic reference for a behaviour does not make it right. In an academic environment, it is no more ethical to purchase work and call it your own to obtain a mark or degree than it is to Photoshop your degree. In an academic context, it is fraud.

Where confusion around this issue might come up is the use of ghost writers and speech writers in a professional work context. For example:

- Political speech writers are paid to write inspirational words for their political masters that the politicians often claim as their own (and for eternity, the pithy quotation will be attributed to the politician, not the person who actually wrote it).
- Advertising reps write testimonials for clients that the client agrees to put its name to, although the company did not write the copy.
- Communications professionals write press releases with quotes from officials that were never actually spoken, but they correctly reflect the official's position on an issue.
- Even in a local government context, reports to the mayor and council are written by staff, but may be released under the signature of the municipal manager (leading some members of the public to believe the municipal manager writes all the reports).

- Individuals with good writing skills may be employed as 'ghost writers' to write the biographies of famous individuals or may be paid as uncredited editors.

In all these situations, work completed by one individual is attributed to another. The difference from plagiarism, however, is that the originator is aware of and is being compensated for this relationship, the public is generally aware that paid professionals were involved in producing the speech or document, *and the work is not being used to obtain a mark or certification*. As a professional, you may produce original work as part of your normal work activities and this work may be used by another. However, do not allow your work to be used for anything beyond what is specified in your terms of employment and by your own ethical standards.

The Ethical Writer

As a student, most discussions on ethical writing focus on the use of sources and citations, and on intentional and non-intentional plagiarism. While these are critically important issues, in both academic and professional environments, a discussion on ethics in writing must encompass a much wider range of issues.

Truth

As an academic or a professional writer, your written word is expected to be truthful (unless it is clearly identified as a work of fiction or creative writing). Beginning with your resume and applying to all written text under your authorship, your professionalism will be evaluated on the accuracy of the information presented.

Avoid embellishment and 'truth stretching'—someone will discover the inaccuracies and the consequences may be devastating to your career. Once in a professional position, your superiors expect that your writing will be well researched with no errors in facts or numbers. While a mistake may sometimes be overlooked, a deliberate attempt to provide false information constitutes 'just cause' and could result in a firing (with no possibility of a reference).

In addition, do not fabricate information. In an academic setting, fabrication could involve intentionally publishing false research results, either by creating results when the experiment was never conducted or by editing or amending obtained results to fit a desired outcome. In a professional environment, fabrication could include the description of an incident that never occurred or creating a falsified 'backstory' to explain an incident. In both academic and professional environments, fabrication is polite terminology for a lie, and when caught, the consequences for the liar may range from charges of academic misconduct to dismissal.

For both professionals and academics, being truthful must also include noting when it would appear that straightforward information is not being provided by others. If the source information does not appear to be correct or the source seems less than honest, be certain to triangulate the information through other sources. Repeating false information (even if another source is referenced) perpetuates the inaccuracy.

Box 5.4 Triangulating Data

Triangulating data (or triangulation) means to consider results from a number of different sources (with a minimum of three, like the points of a triangle) to provide greater confidence in the reliability of the data.

As an example, interviewers at an incident (either the police or the media) may ask a number of witnesses to describe the event. Assuming that there has been no cross-contamination or collusion among the witnesses, the more times an event is described the same, the more likely it is that the event occurred in that manner.

In research, data can be triangulated against other sources of information. If your project involves the pattern of development of a city, sources could be interviews with long-time residents, archival research of photographs, and, potentially, research through historic documents such as the minutes from council meetings or legal agreements. Verification of data from more than one source ensures reliability and adds to the depth of the research.

Be prepared to guarantee your work for accuracy and learn to proofread your work to limit errors. While proofreading is painstaking, the consequences of not doing this carefully can be more so: use a rule of four in proofreading, once for spelling, once for grammar, once to improve meaning, then add the fourth: proofread for accuracy. If there is any possibility for error, particularly in numbers or dates, reference back to the original source and confirm you are correct.

As a professional, you may also want to consider the standards for truth of your organization. If the standards do not match your own, seriously consider if that workplace is the best place for you.

Integrity

The term 'integrity' means abiding by a moral or ethical code in all aspects of academic and professional behaviour. Having integrity means that your work can be trusted and that you perform to ethical standards. Along with personal integrity, many professions have a defined code of conduct that clearly specifies the standards and actions considered acceptable within the profession. Across many professional associations, core elements in codes of conduct generally include:

- to do no harm—not only to patients for medical professionals, but also to any client, test subject, or colleague;
- to respect the values of others;
- to provide true information;
- to act within legal boundaries.

These codes may be simple—as in the 14 points included in the Code of Conduct for the International Association of Canine Professionals that establish the moral compass for the organization—or as complex as the Canadian Bar Association's 172-page document, which sets out acceptable parameters of behaviour and action for its membership. Overall, these codes speak to integrity, that is, performing to high moral and professional standards.

In the social sciences and humanities, an often cited example of questionable research methods is that of sociologist Laud Humphreys, whose methodology in researching for his book, *Tearoom Trade: Impersonal Sex in Public Places* (Chicago: Aldine, 1970), stepped over the ethical line in the eyes of many colleagues. Humphreys posed as a lookout at a public washroom (referred to as a 'tearoom') known to be frequented by men seeking homosexual encounters. Later, he wrote down the licence-plate numbers of the individuals he had observed, obtained their addresses and contact information through the state vehicle registration agency by misrepresenting the research, and then visited each of the individuals at their homes, requesting their participation in a marketing survey. The objective of the research was to determine if the survey responses of the targeted men differed from a random sample of adult males. The academic community's reaction to the research methods was varied—some commended Humphreys for advancing research; others questioned the integrity and ethics of the methods used. This research is frequently noted as one of the reasons why universities across North America instituted research approval committees.

Although you may not have fully articulated it, you likely have your own code of conduct—the strictures you work within, your ethics and morals, and the 'lens' through which you view the world (that is, the way you understand your own behaviour and the actions of others). Take some time to consider your own set of rules shaping your actions. What behaviours do you value? What would constitute a violation of your own moral code?

When you accept a professional position, you agree to follow your company's code of conduct, and in most instances the company's standards will likely mirror the norms of the host society—the generally accepted rules of behaviour and conduct defined both by law and by practice. However, problems can arise if the company's moral code does not match your own. For example, as a professional writer, you may be instructed to produce a document from the organization's viewpoint: if you are employed by a cattle association magazine, you may be asked to write on the superiority of meat proteins contrary to your beliefs as a vegan, or perhaps your right-wing employer will tell you to extol the virtues of 'trickle-down economics', a principle that conflicts directly with your personal socialist-leaning beliefs.

Consider your work from the perspectives of both personal and professional integrity. Hopefully, the majority of assigned professional writing tasks will closely align with your personal standards. If they do not, you may need to consider if you are in the right organization or job. Maintaining your personal and professional integrity in professional writing must be as important as advancing your career.

Responsibility

The porous nature of digital information means that very little written correspondence can be guaranteed confidential. Are you prepared to place all of your writing—papers, e-mails, drafts, and correspondence—up for public scrutiny?

Deleted data can be retrieved by computer techs; e-mails are inadvertently sent to the wrong individual; citizens file access to information requests under federal and provincial legislation to access public documents (called 'freedom of information' laws); correspondence is legitimately or illegitimately placed on the Internet and is available to the world. Your employer can read your blog, or worse, may be reviewing every outgoing e-mail. Does your writing meet your professional standards?

Before you send the e-mail, fax a business letter, or wrap up your daily entry on your blog, consider if you would support the contents being published as the headline story in tomorrow's daily news. Would you like your grandmother or employer to read it? Be prepared to take responsibility for your writing. Do not produce any written documentation that you are not willing to defend both legally and in the court of public opinion.

Chapter Review

This chapter comments on sources and citations, plagiarism, and ethical writing. For sources: cite them correctly. For plagiarism: do not do it. For ethics in writing: know that you will be forever evaluated by your written work. A term paper written in university and saved to a website may be viewed by a future employer 10 years from now. Be certain that every work you write is truthful, accurate, and defendable.

Review Questions and Activities

1. Go to www.famousplagiarists.com to view examples of academic and professional fraud.

2. What three characteristics identify the ethical writer?

3. Take a half-hour to research 10 references in your field of study. Record the reference in APA style in RefWorks or a similar on-line database (whichever is used by your university). Gain familiarity with the referencing system and record all references going forward.

4. What are the four kinds of plagiarism? Which is most likely to be encountered in an academic setting? In a professional setting?

5. Review 10 academic journals in your field of study. Are the articles peer-reviewed? What criteria must be met to have an article considered for publication?

PART II

Research Skills

Part II focuses on the basics of research—how to adapt research approaches to the kinds of projects and experiences likely to be encountered in the workplace. Chapter 6 surveys the necessary research skills, while Chapters 7 and 8 examine research reports and field reports, so that students can take a more comprehensive and holistic approach to understanding and recording observations.

Chapter Six

Research and Knowledge

As a student, most of the writing you do in university in the social sciences and humanities is based on research. You begin with a question (or series of questions), investigate source materials, collect data, complete an analysis, and draw conclusions. This learned skill—the ability to research and record information, then use this information in a written or oral presentation to make recommendations or predictions—is one of the most important skills acquired through post-secondary education. In virtually any professional career, good research skills are invaluable.

Research is a process of discovery, where new information is unearthed, new connections are formed, and knowledge is advanced. Social research has the added dimension of unpredictability, as human subjects do not always provide the responses envisaged by the researcher. This unpredictability does not mean that research is without form and process; instead, just as with research in non-social areas like chemistry and physics, the researcher sets out a framework for the research and follows a logical, systematic, disciplined method to move from the research question to the research findings.

This chapter investigates the sources of knowledge and the basics of research for the social sciences and humanities. The researcher in these fields of investigation has the dual role of being a scientist and of being part of the social world under investigation. These two roles add an additional dimension to the research, as the researcher must always be aware of the biases and preconceived notions he or she brings to any research project. As such, issues surrounding researcher bias are considered, and ways to mitigate (or at least recognize) bias in research are recommended. The chapter begins by examining sources of knowledge, and why we believe what we do in the social sciences and humanities.

Sources of Knowledge

How do you know what you know? Why do you believe it? How is knowledge produced? Much of what we know is not from direct experience. We do not need to stand on the surface of the moon to believe that there is no breathable atmosphere in space, as recognized experts have told us that this is fact. We do not need to rub down with poison ivy to know that the phrase 'leaf of three, leave it be' is worth some credence, as we have learned from others with personal familiarity of species within the *Toxicodendron* genus.

If knowledge is not always strictly experiential, then there must be other sources of knowledge that provide information to shape our perceptions of reality, our belief systems and moral codes, and even the hypotheses we form for research. Three other sources of knowledge—cultural, authoritarian, and superstitious sources—are outlined below, followed by a discussion on experiential knowledge.

Cultural Knowledge

Before mechanization in travel and communication technologies made the world a smaller place, and before knowledge was stored in written works, cultural knowledge was passed verbally and by example to subsequent generations within a society. Children learned from their parents and elders through instruction,

Box 6.1 Gaining Cultural Knowledge

The expansion of economic opportunities due to the digital movement of data, free movement of goods, international trade agreements, and transnational corporations has led to the development of a plethora of websites, consulting groups, and books that provide information on cultural norms among societies across the planet.

For example, the US government publishes specific country studies to be used by citizens and government representatives working abroad. These guides provide information on a wide range of issues, including meeting etiquette, protocols, conversational norms, and ways of showing respect.

In Canada, the federal government publishes guides for people coming to Canada (see 'A newcomer's guide to Canada' on the Citizen and Immigration website, at: www.cic.gc.ca) as well as advisories for Canadians working and travelling abroad (www.voyage.gc.ca). These guides are intended to inform travellers to Canada on Canadian customs and ensure that Canadians travelling overseas do so safely and inoffensively.

Comedians like Rick Mercer (*This Hour Has 22 Minutes* and *The Rick Mercer Report* on CBC) have found humour in the lack of cultural knowledge that others have of Canada. His interviews with residents and political leaders from the US highlight the lack of information even geographically proximate cultures have of each other. Few will ever forget, during the 2000 US presidential election campaign, Mercer asking George W. Bush about future relations with Canada and with Canadian Prime Minister Poutine if he were to be elected, and Bush's straight-faced response.

The benefits of gaining cultural knowledge will ensure that businesses and governments continue to try to understand the differences among cultures—while the motivation may be purely economic, at least it is true that cultural knowledge is growing and being transmitted across an ever-widening audience.

imitation, and example, and systems of beliefs, behaviours, and norms passed through generations within a group.

New members to a group (generally, the new members would be those born into the group, although conquest was another means of establishing a dominant culture) would be instructed on the mores of the society and would shape their actions to conform to these norms: that is, what are considered appropriate or inappropriate behaviours, manners, styles of dress, customs for eating and food choices, folktales, and the complexity of interactions are all defined by the social group through cultural knowledge.

Information could also be obtained by contagious diffusion, requiring a member of a culture to have direct, face-to-face contact with a member of another group to become aware of cultural differences. This individual would then carry information back to his home society, telling others of the experience and allowing them to learn without direct contact. In a western context, exploration and expansionist practices assisted in transferring ideas across the globe.

Today, the processes of globalization and the rapid movement of people and ideas across the planet have increased the cross-pollination of cultural knowledge. The traditions of a culture other than one's own can be assumed either by choice or by acculturation within a new host society. We adopt styles of dress, cuisine, and words from other languages at an increasing pace—one does not have to be from Scotland to wear a kilt, from India to make samosas, or from Japan to enjoy sushi.

The development of communication technologies has shattered the separations among cultural groups. Where technology is available, the usual boundaries lose meaning: an international boundary has little meaning to someone with full access to the Internet, a cellphone, and a valid passport. New opportunities now exist for individuals to learn about other cultures and define their own cultural mores. Students and professionals have far-ranging opportunities to learn about other cultures, both through written documentation and personal experiences.

Of concern to some researchers is the loss of cultural knowledge as the distinctions among cultures become blurred due to the transnational availability of consumer goods and information. In particular, the loss of language is cited as a concern as English becomes the worldwide lingua franca of business and technology. Linguists anticipate that the number of languages in the world will be reduced from a current 6,000 to about 600 over the next few generations, and language, as the Québécois and other groups in Canada have taught us, is the core element of culture.[1] The impact of culturally gained knowledge on an individual's belief systems appears to be declining: authoritarian knowledge is replacing knowledge transmitted within cultural groups as the most relied upon source for information.

Authoritarian Knowledge

If a recognized authority provides information and this information is accepted as true, then the authority has contributed to the respondent's base of knowledge. Experts may be parents, professors, or peers—those with whom the respondent has direct contact—or the authority may be a person far removed from the learner, with contact through text or other media.

This external authority may be an individual (a sports hero who is viewed as a 'role model'), an NGO (non-governmental organization) such as Greenpeace or Amnesty International, or a reputable source (for example, Statistics Canada)—in any case, the individual endorses the knowledge passed on by the expert and accepts it as true. The respondent gives sanction to (that is, gives authorization to) the experts, allowing them to create new beliefs for the respondent.

For example, parents have enormous authority in shaping the beliefs of children (with reference to Santa Claus, parents can create beliefs that are not precisely true along with fact-based beliefs), but as the children grow and expand their range of influences, new authorities are recognized. As teens, they may rely on their peers to determine everything from fashion to career choices. Celebrities and media personalities may be viewed as authorities (although it can be questioned if the celebrity has actual knowledge in the subject area or not). With age and experience, fields of contact grow through direct interactions and formal and informal learning. New authorities are constantly identified. Throughout a human's lifespan, we choose authorities that we believe in, give sanction to, and trust, and these authorities shape our belief systems.

Hopefully, the trust in the authority is well placed. From the earliest of human societies, experts have espoused facts that later have proved to be false: until only a few hundred years ago, most learned authorities thought the world to be flat and that disease was passed among populations by mysterious miasmas. While misplaced trust in the knowledge of authorities may seem harmless, the impacts can be far-reaching. For example, during the first quarter of the twentieth century, social scientists were enamoured with *environmental determinism* as a belief system, and the ideas behind this theory shaped much of the writing and opinions of experts. Leading theorists like Fredrich Ratzel, Ellen Semple, and Ellsworth Huntington all espoused theories that human capacity was determined by the environment—you are the way you are based on where you are from. That is, human behaviour, intellect, and capacity are shaped by the person's location on the earth's surface (at a time when populations were less mobile than they are today). The theory was considered immutable and many believed that all human behaviours could be understood in relation to the individual's place of origin. Relatedly, theories of social evolution—that all societies and individuals must necessarily evolve along a similar course from 'savagery' to 'civilization'—conveniently placed the white, male, Western theorists at the top of this evolutionary chain. The results of such 'knowledge' could be seen in the warfare and colonialism of the nineteenth and twentieth centuries, and continue into the present.

While these social theories no longer hold academic credence, the belief system upon which they are based continues to shape conflicts among cultural groups across the globe. Knowledge transmitted by authorities and believed by members within a cultural group is at the root of many current hostilities. The danger of giving sanction or trusting an expert opinion can be highly damaging.

Across the planet, we are bombarded with expert opinions daily from a wide variety of sources: print advertising, television, movies, text messages, and the Internet provide up-to-the-minute information on what is known about what to wear, eat, and drink, how to behave, and what to desire. Advertisers add legitimacy to ads by presenting them as news stories (infomercials), and newscasts are presented as entertainment to gain higher ratings, and, therefore, more advertising dollars. Celebrity spokespeople assume the role of expert to tell listeners what products to acquire or what not to do. It might be fair to ask, for example, if ex-Beatle Paul McCartney knows very much at all about the historical, cultural, and economic significance of the Canadian east coast seal hunt, yet that does not prevent him from gaining 'photo ops' in his efforts to end the hunt.

As a receiver of information, the recipient must carefully filter through all available data to determine what is legitimate knowledge and what holds little basis in fact. In addition, the research methods used to obtain the information should be examined and determined to be reliable, and the study itself should be sufficient to be able to produce an expert opinion before any trust is placed in the information.

From both academic and professional perspectives, the credentials of the authority are also important, as is the person's area of expertise: a Ph.D. in biology does not necessarily qualify the individual to speak with authority on nuclear physics. The recipient of 'knowledge' must be cautious of giving sanction to 'experts' without being certain of their qualifications and research methods. A critical mind is the best weapon in evaluating knowledge gained from experts in any field of study or professional career.

Practical tip. Take some time to consider who and what you believe in. What are your own sources and references? Upon examination, are they valid?

Superstitious or Stereotyped Knowledge

If a knowledge statement begins with 'everyone knows that . . .', it is possible that the speaker is working from knowledge based in superstition or stereotypes.

For the most part, superstitious beliefs are innocuous and add colour to our everyday experiences. In Western cultures, we wish on four-leaf clovers and hang horseshoes over doorways for luck: the person taking action does so to perpetuate a cultural belief. Superstitions are found across cultures worldwide, and exchanges in cultural knowledge are increasing our awareness of these beliefs. For example, the cultural dislike of the number '4' has been imported from Asian cultures into the West, as seen in addressing and hotel room numbers in venues that cater to expanding tourism markets. The favouritism shown to the number '8' is also seen in the names of retail venues and telephone numbers. The superstitious beliefs are recognized to show cultural respect and, more practically, to ensure that the potential negative economic impacts are decreased should a business be considered suspect by the target market.

Box 6.2 Critical Thinking

Richard Paul and Linda Elder, in *The Miniature Guide to Critical Thinking: Concepts and Tools* (2006), note that critical thinking does not follow a rote process, but instead involves obtaining information, then evaluating the veracity of this information by examining the information from several perspectives. That is, the critical thinker uses learned skills to not just be a tacit recipient of information, but to be an active learner. The perspectives of a critical thinker include:

- evaluating one's own biases or prejudices towards the information and determining if these are shaping one's interpretation;
- considering the source of the information, and whether or not the informant is qualified to make a statement or comment;
- using logic and rational thinking to evaluate information;
- evaluating whether information presented can be substantiated in other reputable sources (placing the source within the context of information available in that field of study);
- recognizing assumptions;
- taking an open-minded approach to new information;
- suspending judgement until a wider perspective can be obtained (and avoiding snap judgements);
- evaluating arguments or lines of reasoning;
- conceptualizing relationships among pieces of information to obtain a wider view.

While taking a critical-thinking approach will not always guarantee that a correct conclusion has been reached, at least the learner can be assured that he or she is taking an involved approach to discerning the value of new information.

Most superstitious knowledge, when dissected from a logical perspective, is difficult to defend. If it were irrefutably true that 'bad luck comes in threes', then we might have to accept that all outcomes are predetermined and that humans must tacitly accept that one perceived negative event will automatically trigger two more in short order. If it were true that finding a penny guaranteed a lucky day, then it would make sense to strategically place pennies where they could be found by the penny placer or persons for whom the placer wished to cause good luck.

While most beliefs are more culturally interesting than harmful, more damaging is the acceptance of ethnic or gender-based stereotypes. These generalities can limit the recipient's ability to accurately understand a situation or create negative reactions to other individuals.

There may be elements of truth in superstitious knowledge (it does make sense not to walk under a ladder, as this will decrease the likelihood of being coated by a bucket of paint or hit by a falling hammer), but care must be taken to ensure that this form of knowledge is not limiting, inaccurate, or patently false. As a student or professional, it is important to consider the source of knowledge. Use critical thinking skills to discern if information is actually true or if it is based on inaccurate information or incorrect assumptions.

Experiential Knowledge

A portion of our knowledge also develops from direct experience. Our first-hand discoveries and experiences are the most powerful in shaping our beliefs and actions. We know that touching a hot surface will burn our hand because we have directly experienced it, just as we know that there is a correlation between completing class assignments and the likelihood of a favourable mark.

While experiential knowledge has the advantage of being unfiltered information, errors or omissions can occur. For example, an individual may generalize his or her experience to apply to all similar situations—one turbulent flight may cause the individual to classify all air travel as dangerous. As humans, we may also use past experiences to allow us to come to rapid decisions and conclusions on new experiences. Ideally, we can evaluate our experiential knowledge in the same manner as we evaluate other forms of knowledge—by being aware of faulty connections among events or misplaced logic in our interpretations.

Experiential knowledge may be gained informally through day-to-day events that cause us to gain new knowledge or understanding, or formally through experimentation. In the latter, the experimenter develops a research question, and then manipulates the research environment through testing or other methods to see if the question can be answered.

With both informal and formal experiential knowledge, the researcher must carefully examine her or his own approach to obtaining and understanding knowledge. Creating a research question and methods to prove a theory is a well-tested tradition in science, but both the question and the methods can be manipulated to achieve a desired conclusion. The next chapter examines the basics of research so you can better understand the relationships among beliefs, accuracy, and the role of the researcher.

Being aware of the sources of knowledge is important in the evaluation of the veracity of information. In the academic world, a researcher will seek to understand how knowledge has been formed and whether or not it is valid, and whether the researcher's own biases and beliefs are shaping the collection of data or the interpretation of results. In a professional milieu, it remains critically important (with financial or career implications) to know the source of information, how information was obtained, and whether or not it can be relied on as fact.

Chapter Review

This chapter has examined sources of knowledge—cultural, authoritarian, super-stitious, and experiential. It is important to note that classifying sources of know-ledge is in no way meant to suggest that one source is superior to all others. We can falsely interpret a first-hand experience just as we can buy into a late-night infomercial populated by fake authorities with counterfeit credentials. More important is the student's and professional's ability to investigate the source of information and use critical thinking skills to evaluate contentions, conclusions, facts, figures, and statistics.

Review Questions and Activities

1. What is the difference between authoritarian knowledge and experiential knowledge?

2. Consider any superstitions you may hold to be true. Using a critical-thinking perspective, what basis is there in fact for these superstitions?

3. Select from a recognized publication (either an academic, peer-reviewed journal or a piece from a major newspaper or news magazine) an arti-cle and review it from a critical-thinking perspective. Are sources doc-umented? Are the methods transparent and replicable? What is your reaction to the writing?

4. What authorities do you recognize in your field of study?

5. How does culturally obtained knowledge shape your belief system?

Chapter Seven

Research Basics

Why do research? In academic and professional contexts, most of the research we do is intended to fulfill one of two tasks:

1. to prove why something happened in the past;
2. to show what will happen in the future.

Generally, we are seeking *causality*—the link between cause and effect that shows how one event or action has a direct and usual influence on a subsequent event or action. We seek out the relationships to understand the past or to give us the ability to predict what will happen in the future. If causality can be firmly linked and widely believed, we may elevate the relationship to a postulate or law—a relationship that appears to be true all of the time.

This binary-pairs approach (cause and effect) shapes much of research based in Western scientific traditions. The researcher sees an effect and attempts to find the cause by taking a logical, rational approach based not on bias or assertions but on *science* (more on this to follow). Within this approach, researchers generally take on one of two kinds of arguments: inductive or deductive. *Inductive research* moves from the specific to the general. A generalized research finding based on experimentation or observation uses inductive logic. *Deductive research* moves from the general to the specific. A research finding that is based on laws or widely accepted principles uses deductive logic.

In the social sciences, research from an inductive approach attempts to create theories from surveys, experiments, case studies, and observed examples. That is, if it is true for case study after case study, then the researcher may be able to develop a generalized statement or law about the research. As long as the research is based on a rational premise, carefully conducted, and follows a logical progression, then a conclusion with widespread applicability should develop. For example: *If Tim is a student, and Tim drinks coffee, then all students likely drink coffee.* The researcher may conclude that an investment in an on-campus coffee shop would be a wise financial move. The researcher *infers* this larger conclusion from the specific examples observed of Tim and his colleagues drinking coffee.

As a scientist, you are likely already thinking of the danger inherent in an inductive approach: the individual (as a research subject) does not always behave in a manner that leads to a common conclusion, and myriad variables may impact the results. Drawing a conclusion that is true for all people in all instances can be difficult.[1]

Research from a deductive approach sets out a premise *known* to be true and then proves the veracity of the premise through experimentation. The research concludes by showing that the experimentation supports the original premise. That is: *If all students drink coffee, and Tim is a student, then Tim drinks coffee.* Tim's individual coffee habit enforces the initial, widely accepted premise: that all students drink coffee.

A deductive approach requires that the conclusion must follow logically from the premise and prove the premise to be true. Much of scientific research proceeds from this approach, where a hypothesis is crafted and the researcher sets out to determine if the original premise is verifiable. Again, there are difficulties with this approach: if the original premise is false (whether this falsehood is known or unknown to the researcher), then the research is working to verify a result that is incorrect. In addition, if the research does not provide the required result, the researcher may think that the experimentation has been done incorrectly and adjust the research methods to obtain the 'correct' result.

Consider the following example:

> Tim: I have observed that that cafeteria serves turkey tetrazzini every day after we have turkey burgers. Since we had turkey burgers today, we will have turkey tetrazzini tomorrow. This pattern seems to always be followed.
>
> Anna: That's Cafeteria Law. All leftovers must be reused in a different format.

Tim is using an inductive approach, moving from specific observations to a generalization, while Anna is working from the opposite perspective, citing a general law (or widely accepted belief) and applying it to the specific example of the cafeteria menu for tomorrow. It is noted that both students are speaking to the same phenomena, yet each takes a different approach.

For any research question, with some effort in logic, an argument may be shaped from an inductive or a deductive approach. The difference is in the research methods that will be used: Tim will provide additional case studies as evidence of his conclusion, while Anna would discuss the rules of food usage in cafeterias and cite long-standing business cases that support the reuse of food.

As with most dichotomies, it is possible to combine aspects of the two into a merged approach to research. That is, a researcher taking a deductive approach might also start by examining case studies, and a researcher using inductive methods might consider what is known or widely accepted to be true in the particular field of study. In research, both widely accepted principles and collected data are important, and disagreement over which is a better research approach often ends at a 'chicken and egg' argument: grand theories cannot be crafted without research, and research benefits from building on grand theories.

These two approaches (or a combination) are used to elevate simple explanations or reports of phenomena in an attempt to understand the phenomena: it is this shift that leads to scientifically based research.

Doing Research

What is research? Research is different from a casual conversation on causality because it requires two additional and elevating factors: the ability to replicate results and recording of the methods used to obtain the results. That is, research is more than just making a statement on what one believes to be true or false: through experimentation or the application of research methods, the researcher attempts to learn if an idea (or hypothesis) has scientific merit.

Research starts by asking questions. The six foundational questions are:

- Who?
- What?
- Where?
- When?
- Why?
- How?

Box 7.1 Serendipity and Sagacity

'Serendipity' is defined as accidentally discovering something new and important. Related to this discovery is 'sagacity', or the ability to link seemingly unrelated factors to valuable conclusions. Every student is familiar with Isaac Newton's 'discovery' of gravity from watching an apple fall from a tree. A more prosaic example is Post-it Notes, developed by scientists at 3M. At first, a glue that was not very sticky appeared to have no useful purpose. However, 10 years after its invention, a scientist remembered the product and used it on small pieces of paper to mark pages in a book. The usefulness remained elusive; it was only after giving out the product for free that consumers began to be interested in this now ubiquitous product.

In other words, should unexpected results be obtained by the researcher, the researcher must be able to step back from the research question and see if the results answer another (but no less useful) question. Sociologist Robert Merton examined serendipity in social science research and speaks to the 'serendipity pattern' as the fairly common occurrence of unanticipated results that then leads the researcher in new directions.[2]

Therefore, findings that do not agree with a research question may themselves be worthy of further research. As stated by Isaac Asmiov: 'The most exciting phrase to hear in science, the one that heralds new discoveries, is not "Eureka!", but "That's funny."'

While this might seem to be an oversimplification of the depth and breadth of research in the social sciences, most research questions can be broken down into a variation on these six questions:

- Who is involved?
- What happens when A and B interact?
- Does place have an impact on the results?
- Does time?
- Why is this happening?
- How is it happening?

Research in the social sciences is fascinating work. Social scientists investigate human subjects, and may consider a wide range of human relationships:

- *Roles*: the part played by each human actor at work, at home, or within any of the above relationships;
- *Relationships between individuals*: parent/child relationships, for example, or superior/subordinate relationships;
- *Relationships within a group*: for instance, cliques at a high school or interactions at a workplace;
- *Relationships among groups*: observations on larger populations, such as income or crime distribution in a city, the formation of neighbourhoods, ethnic distributions;
- *Subgroups*: bull riders, skateboarders, church communities, or sociologists, for example.

In the study of human subjects, the researcher seeks to understand events in their natural setting, then report on them from a scientist's viewpoint. Critical to this role is that the researcher must understand that he or she has preconceived biases and perceptions on the research, and manipulation of the research process and interpretation of results might follow from researcher bias. In some disciplines (notably anthropology) and from some ideological/methodological perspectives (notably feminism), researchers often make a conscious effort in the course of research and in their reports on the research to note their own biases and to discuss the effects their presence had on the research and its subjects, and on its analysis. This is commonly called *reflexivity*.[3] While there are myriad roles for the researcher, four of the most common are described below:

1. *Complete observer*. There is no interaction with the human subjects. Ideally, the subjects will be unaware that they are being observed, as people may behave differently if they feel they are being watched. The researcher maintains a distance from the subjects and attempts, through observation, to understand the interactions and events.

2. *Participant observer.* Participation may occur at different levels, but the key is that the human subjects are aware of the presence and role of the researcher. The researcher may be a passive participant, taking notes and observing, or take a more active role as an adopted member of a group. Again, human subjects may alter their behaviour if they know they are under observation.

3. *Covert participant.* By far the most difficult role for the researcher, in this role the researcher will assimilate fully with the human subjects as one of them. The researcher does not identify herself or himself, and any notes are completed after the researcher has left the research setting. While covert participation may yield the most accurate (or interesting) data because participants are acting in their normal state, it requires extended time frames to find a means to successfully access the group, and the researcher may lose perspective if the assimilation is deeply developed. In addition, the betrayal experienced by the subjects when they discover that they have been unwitting participants in a social experiment may be devastating, and potentially ethically indefensible.

4. *Non-participant in on-the-ground research.* The researcher is not 'on-site' as part of the research. Instead, non-contact methods of research are used, such as questionnaires sent to 'participants' selected at random, archival research, or compilation of secondary source data. The researcher has no direct contact with the research subjects.

A review of academic journals in your subject area will be invaluable in gaining information on the methods used by other researchers. Read through current issues of journals to understand the methods being used by other researchers. Are there trends in research in your field of study? For example, historians may focus on archival research, while cognitive psychologists look to experimentation to obtain data. Knowledge of a wide range of research approaches will allow the researcher to vary her or his methods to fit different research questions.

Who Does Research?

We all do research, whether in shopping for groceries, buying a car, looking for a new sound system, or searching the Internet to find out what we can about a musician we just heard on the radio. At times, we will pursue this research with some degree of care and order, at which point we are beginning to take a scientific approach to our curiosity or what we want to find out. Identifying oneself as a scientist or identifying research as scientific implies that there is a certain rigour in the formation of the research question, that the methods used are valid and reliable, that a systematic approach is taken to collecting data, and that the research results are reported without bias or manipulation.

In casual conversation, we might make observations from an inductive or deductive approach, and may even use some rigour in our arguments, citing case studies and examples or referencing back to widely accepted principles and laws. However, the listener has different expectations for the speaker in a casual conversation than the scientific community (and the public) has for someone portraying him/herself to be a scientist. For scientific research, we have certain expectations that include the following:

- *Scientific research is unbiased.* We expect that a scientist will approach a research question from a logical, analytical perspective. Even though it is known that scientists carry with them their own understanding of the world, based on their own experiences and biases, we expect that a 'real' scientist will be able to step back from these experiences and biases to take a rational approach to developing research and interpreting results. We expect that social scientists will understand the concept of reflexivity, realizing that they are people researching other people and hence are part of their own research. Social science has struggled against this dialectic (of the researcher being part of the research by virtue of being a human); however, this in no way lessens the importance of social research. Instead, additional attention must be paid by social researchers to understanding their own personal biases, their emotions that shape research decisions, and their motives in verifying results.
- *Data collected in a scientific manner is valid.* The use of statistics implies that a certain level of rigour has been used in data collection, that the results are true, reliable, and valid, but this is not necessarily the case. Just because the data showed the desired result does not mean that they are reliable: the sample used may not be sufficient or may be selected to achieve a desired outcome; the questions in a survey may be formatted in a way that encourages particular answers; a questionnaire may have been distributed to a narrow segment of the population or with no consideration as to who will and will not respond or participate; the results may be only partially published to support a conclusion.

Practical tip. In the future, pay particular attention to the methods used in data collection when reviewing any studies. Readers often pass over the methods section in a research report or article and jump to the conclusions (the interesting part of the study). However, the way the data were collected (and then interpreted) is critically important. Often, careful review of a researcher's methodology will raise questions for the reader and may cause the reader to re-evaluate the effectiveness or accuracy of the study.

- *Scientists use a system and report their results.* We expect that a true scientific study will identify the research methods, show how the research was conducted, and present results in a way that anyone interested in putting out the effort could replicate the results. This requires that a scientist publishes the research results and accepts the criticisms of peers should the research be found to be less than scientific in format or conclusions. It also assumes that research focused on humans is necessarily replicable and, often, that the conclusions drawn from research are applicable to all people. But groups of people and individual persons are unique, and no social science fieldwork can be precisely replicated. Likewise, medical researchers over several generations in Western society focused most of their work on white males and assumed, often wrongly, that their findings applied equally to women and to other, often racialized, groups.[4]

- *Scientists are progressive and 'truth-seeking'.* Above all, we expect that good science is about finding truths, working forward from what is known to what is unknown, filling gaps, and advancing human knowledge. There is little point in replicating the same study over and over if no new information can be provided; imagine a university where every researcher worked on the same study and taught the same course. This assumption, however, is categorically false—not all scientists (including social scientists) are progressive: some are conservative, others reactionary; and, inevitably, some seek only the truths they want to find and shape or analyze their data to suit these 'truths'. While questions of bias and misreporting of data will always be part of science, the search for truth and making a real contribution to knowledge remain the core purpose of 'science'.

How to Conduct Research

Research starts by asking a question, generally stated in the form of a hypothesis: a simple statement of a proposition, but without any assumption that this proposition is necessarily true. The research to follow aims to test the truth of the hypothesis.

In proceeding to test the hypothesis, the researcher must consider personal bias, first, in the shaping of the research question (that is, is only one possible outcome being considered or allowed because of previous experiences or knowledge?). The researcher also must consider if the selection of particular research methods will predetermine that the hypothesis is true. A researcher inevitably carries his or her own ideas and previous experiences into the research—beliefs about social classes and 'others' perceived as unlike the researcher. The researcher must carefully consider if the research question is shaped to achieve only one possible outcome. The consideration of bias and approach is critical: a good scientist will spend a great deal of time developing a good research question, one that is clear, pointed, unbiased (as much as possible), and that builds on previous research.

A hypothesis is a working assumption. The scientist develops an idea, and then designs a research program to see if this idea has scientific merit. The scientist will look to previous research related to the hypothesis, and then will craft a research program (experiments and observation) that will test the veracity of the question. Ideally, the research should achieve expected results that prove the research question to be true. However, proving that a question is false or obtaining a different set of results from what was expected is not always negative: these latter results could take an open-minded researcher in new, unanticipated directions (see Box 7.1).

When crafting your hypothesis, limit your topic to a single purpose statement that you can easily explain in one or two sentences. If you do not have clarity around your hypothesis, the rest of your study will lack focus, and valuable time will be wasted as you attempt to construct a research program without a full understanding of the research topic. For example, perhaps your primary interest is shopping. You decide to make shopping the topic of your senior research paper. In limiting your topic, you realize that your interest is in the newest form of retailing, the lifestyle commercial centres that are evolving on the retail landscape. You wonder how these new forms of retailing are developing, and what mechanisms are in place to permit their development. You suspect that slow growth of these centres must be due to prohibitive zoning and land-use regulations. Therefore, your hypothesis will focus on the zoning and land-use regulations that contribute to or prohibit the development of lifestyle commercial centres. Instead of investigating the social, economic, or environmental aspects of these lifestyle commercial centres (which all are valid research questions), your research program develops around understanding both the distribution of these centres and the means by which they are sited. As you proceed, you will investigate case studies where lifestyle commercial centres have been successfully sited as well as instances where these developments have been refused by local governments. Ultimately, you will develop an understanding of the factors that contribute to siting these centres that could be applied to proposed developments in the future.

Practical tip. When embarking on a major project (or thesis or dissertation), take some time to craft your purpose statement. You should be able to explain your research in plain language so that any interested person can clearly understand the point of your work. You may feel that your research is just too complex for the average person to understand: this is a mistake. While not everyone will fully comprehend (or be interested in) the intricacies of your multivariate statistical analysis, it is incumbent on you, the researcher, to be able to state clearly the purpose of your work. Time spent on clarifying the research question will equal time saved in the research.

A hypothesis sets out the structure for the research. It is a means of focusing and limiting the research to an answerable question. Ideally, the hypothesis is limited enough to focus the research but crafted with depth to ensure that the research adds to the body of knowledge on the topic.

A problem can arise when a researcher chooses a question that cannot be researched: when the application of research methods to the question cannot lead to any sort of resolution or conclusion. Some questions, after all, have no real answers:

1. Is purple better than blue?
2. Is capital punishment always more effective than talk therapy?
3. What is the best way for all students to study?

The first question is one of subjective preference: there is no real scientific answer to the question, although individuals may have strong opinions on the topic. The second question is not answerable. It appears to compare two responses to crime and criminals, but the two responses are unrelated. They speak to a value judgement on the part of the researcher and the reader, and potentially to legal issues, but empirical research would not be able to prove that one is a more effective response as set out in this hypothesis. The third question, again, cannot lead to a 'right' answer for all students in all situations. The best method of studying depends on the student and the material being studied, as has been shown in previous research. While it is scientifically valid to question the findings of previous research, ideally the researcher is building research forward, not asking a question that is unanswerable.

The effort taken to craft a hypothesis elevates a question to the level of science. The researcher is, in effect, declaring an intent to take a scientific approach to the research and is willing to put the research to the test of peer scrutiny through publication of the research and results. A good research question should be specific, limited to one or two ideas, have a quality that can be measured or somehow analyzed that allows it to be empirically tested, and have some relationship to previous research to allow for comparisons or verification. Ideally, the research question will be important, practical, and contribute to knowledge in the field of study.

Box 7.2 Occam's Razor

Occam's razor is an approach to research proposed in the fourteenth century by William of Occam, a medieval philosopher and theologian:

Pluralitas non est ponenda sine necessitate.

which translates as 'entities should not be multiplied unnecessarily.' His approach is often incorrectly translated as 'keep it simple', but this loses the core meaning of Occam's approach.

Occam proposed a logical approach to the development of hypotheses. The principle is that, when there are two or more potential theories to explain an event or phenomenon, and both are equally possible, it is more likely that the less complex theory is the correct one.

Thus, if your cellphone is missing, it may be that you left it in one of the places frequented since you last used your phone. It is also possible that your cellphone was stolen from your pocket by a skilled thief and is being used to conduct international crime. A third possibility is that the phone companies have found a way to vaporize cellphones if the user has not paid a bill within a specified time period. In short, your cellphone has ceased to exist.

While the latter two theories on the missing cellphone are possible, they require additional actors and a more complicated approach to finding the cause-and-effect relationship. The first is the most simple, with the fewest steps and fewest actors, and therefore the most likely. The use of logic 'razors' out the other theories: the first theory is investigated until it is proven wrong. If there was other evidence (such as cellphone charges made while the phone was missing, or a notice from the cell provider) then the more complicated theories could be true.

In brief, then, it is not useful to unnecessarily complicate an approach to understanding a problem: the simplest solution, with the fewest steps and actors and the fewest built-in assumptions, is more likely to be correct.

Quantitative and Qualitative Approaches

How do social scientists do research? Traditionally, research is divided into quantitative and qualitative perspectives. This dichotomy is one more of convenience than complete accuracy: most research in the social sciences will contain both quantitative and qualitative perspectives, as will much of the work done in a professional environment. However, this two-pronged classification system allows for a discussion on a range of research techniques and approaches, and each is described below.

Quantitative Approaches

From the early part of the twentieth century to about the 1960s, much of research in the social sciences was couched in quantitative perspectives, based on methods of experimentation carried forward from the natural sciences. Social scientists looked to their fellow natural scientists and attempted to replicate the latter's ability to be seemingly separate from their research subjects, taking an unbiased approach to the development of hypotheses and the conduct of research.

The natural scientists were also admired for the search for laws and order in the universe. Social scientists attempted to replicate this objective approach, with detachment from subjects and the formation of research questions that sought what sociologist Emile Durkheim referred to as 'social facts'.[5] The researcher is able to approach social phenomena from a non-biased approach, setting aside preconceived ideas and searching for those factors that shaped behaviours and events. The social facts are external to an individual, and regardless of the social scientist's opinion or hypothesis, the research will prove out these social facts.

Thus, the quantitative researcher looks for causal factors external to the individual (like population density or socio-demographic characteristics that produce results for an individual, just as the biologist would look to food supply, competitor populations, and climate (and potentially myriad other factors) to understand the rise and fall in frog populations. In purely quantitative research, the scientist looks to take a realist's approach to understanding a situation or phenomena, believing that experimentation and research will lead to an understanding of the independent factors that cause an effect on dependent factors, and ultimately that laws or rules can be found that govern these relationships. At its most basic, quantitative research would examine only factors that are observable and can be measured.

Qualitative Approaches

Since the 1960s, qualitative approaches have gained greater legitimacy in the social sciences as part of a larger movement on a societal level to question laws and authority and to better understand the role of the individual in societal fabric. When taking a qualitative perspective, the researchers recognize that they are actors within a social construct and cannot ever fully remove themselves from this construct. Both the researchers and the test subjects bring their own ideas, history, and agendas to the research, and these perspectives can have a bearing on the findings of the research.

Where the quantitative researchers of the past emphasized researcher objectivity, qualitative researchers recognized that subjectivity had a large role in shaping research questions, methods of analysis, and the interpretation of results. Blanket assertions can't always be made: instead, the results of research might be specific to that situation or research structure, and should be considered as adding to a body of knowledge rather than as a means of uncovering the hidden laws shaping the relationships between causes and effects. Researchers steeped in qualitative approaches have also noted that science itself is a social product, and therefore shaped by the belief systems and values of researchers and the larger society. As noted by Palys, 'There is no such thing as immaculate perception.'[6]

It should be established, however, that research based in qualitative methods is still 'science'—although the purpose of the research may not be to produce grand theories, the examination of individual and group behaviour and the accurate reporting of results add to the body of knowledge known on a particular topic. The researcher's bias, the subject's response, and the context of the research all become part of the research question. How people (the researcher, the subject, and the reader of the study) think will influence how they act and their interpretation of events. The emphasis, then, is on understanding the reality of the researcher, subject, and reader, and factoring this into developing research conclusions or understandings.

Much has been written on the historic context of these two different approaches, and this history certainly is worthy of further study.[7] This brief overview has been presented to set out a context for the next section—the methods used by researchers to understand the world around them.

Box 7.3 Dangerous Research

Even social scientists face danger. David Calvey's research with a group of bouncers in Manchester, England highlights the dangers that can be faced by social scientists. As a covert participant (that is, conducting the research by infiltrating the study group without the knowledge of the bouncers, bar owners, or patrons), Calvey dealt with drunken patrons, gang violence, drug culture, and bar fights in his time spent as one of the group. As a trained martial artist, the researcher had the ability to credibly perform as a bouncer. This may not be true for all researchers: if you are conducting dangerous social research, care must be taken to avoid situations that could cause physical or emotional harm.

For more on social science risks, see G. Lee-Treeweek and S. Linkogle, eds, *Danger in the Field*. The text provides an interesting analysis of danger faced in qualitative social research in field settings including a hospital, police department, seniors' care home, and various urban settings.

Research Methods

As previously noted, most research is an attempt to find out what happened or to predict what might happen in the future. Unless the researcher has the luxury of unlimited funding, most research is purposeful, either to complete course or degree requirements or because it is part of the professional's duties and activities.

The most defining feature of research (both qualitative and quantitative) is that scientists generally use a method to obtain results, and the method used is an important part of the research. Why is the use of a method so important? Because it allows other researchers to:

- understand how the research was done;
- see if any steps have been missed;
- discover if researcher bias is a critical element in shaping the results;
- evaluate the results to determine if they make sense.

The method may be simple observation and recording of a single occurrence of an event, or it may be a random sample trial conducted over many years and many populations. The important part is recording the method used and reporting on the results obtained.

Quantitative research methods most often involve statistics, numerical measurements, and the categorization of data, while qualitative methods are used to describe a phenomenon. Research that employs experiments, statistically significant random sample surveys, and standardized testing involves quantitative research methods. Qualitative researchers use case studies, in-depth interviews, archival research, observation, participation, and ethnographies (the systematic and scientific study of human cultures). In the latter, the attempt is made to understand the situation from the perspective of the human characters, while in the former the attempt is made to elevate the research above the individual's idiosyncrasies.

For example, the following study uses a quantitative approach to understanding how non-verbal cues can increase the persuasiveness of videos viewed by the participants:

> Joseph Cesario and E. Tory Higgins, 'Making message recipients "feel right": How nonverbal cues can increase persuasion', *Psychological Science* 19, 5 (2008): 415–20.
>
> *(Excerpted from the study with the permission of the authors.)*
>
> The researchers used regulatory-fit theory as a framework for understanding the effect of nonverbal cues on a message's effectiveness, and as a foundation for developing a new persuasion technique. They proposed that when the nonverbal cues of a message source sustain the motivational orientation of the recipient, the recipient experiences regulatory fit and feels right, and that this experience influences the message's effectiveness.
>
> The researchers tested two predictions. First, they predicted that regulatory fit produced by nonverbal cues would result in greater message

effectiveness, with the eager delivery style being more effective for promo-tion-focus than for prevention-focus recipients, and the vigilant delivery style being more effective for prevention-focus than for promotion-focus recipients. Second, they predicted that regulatory fit would result in greater experiences of feeling right, and that greater experiences of feeling right would be associated with greater message effectiveness.

Ninety students participated in return for $5 or course credit. They com-pleted the Regulatory Focus Questionnaire and were randomly assigned to watch one of two videos. The two videos resembled professional advo-cacy videos. In both videos, a message source (ostensibly a public-school teacher) advocated implementing a new after-school assistance program for children. The content of the message was identical in the two videos; the only difference was whether an eager or vigilant nonverbal delivery style was used by the source when delivering the message. The participants were then given the questionnaire that asked them to indicate their attitudes toward the program (i.e., how favorably they felt toward it, how good an idea they thought it was, their overall attitude toward it), the extent to which they agreed that the program should be implemented, and their behavioral intentions toward it (the likelihood that they would vote in favor of it). Ratings were made on 9-point scales. A message-effectiveness score was computed for each participant by averaging responses across these items; higher scores indicate greater effectiveness.

Experimental results support these predictions. Participants experiencing regulatory fit (promotion-focus participants viewing messages delivered in an eager nonverbal style, prevention-focus participants viewing messages delivered in a vigilant nonverbal style) had more positive attitudes toward a message's topic and greater intentions to behave in accordance with its rec-ommendation than did participants experiencing nonfit. Feeling right was also greater for participants experiencing fit than for those experiencing nonfit and was associated with greater message effectiveness. Regulatory-fit theory provides a framework for making precise predictions about when and for whom a nonverbal cue will affect persuasion.

For comparison, a qualitative approach was taken by the researcher on this study investigating who gets tattoos (and why):

F.J. Johnson, 'Tattooing: Mind, body, and spirit: The inner essence of the art', *Sociological Viewpoints* 23 (2007): 45–52.

(Excerpted from the study with the permission of the author.)

This research began to understand why people choose to get tattoos. The reason was to find out if getting a tattoo was a novelty or if there was more to it than just what we can see inked on their skin. The interest of this research lies in the feeling, emotion, human awareness of expression, and the deeper meaning on the inside that coincides with what is seen only as skin deep on the outside.

Data were collected through interviews with thirteen people that have tattoos, a sample consisting of seven women and six men ranging in age from 20–65 years old. Four tattoo artists, two men and two women were interviewed. One evening was spent 'hanging out' in a tattoo studio observing and interacting with the patrons that visited. Several conversations were had, which did not constitute an interview as such, with students on campus and other people that happened to come in contact with the researcher elsewhere. The researcher designed and had her own tattoo done.

Five of the people were interviewed in person; the other seven were interviewed online by computer. Face-to-face interviews consisting of both closed and open-ended questions were tape recorded and conducted in tattoo studios. The researcher's interest in the subject and her desire to get a tattoo seemed to make for a deep connection with those that she interviewed.

The reasons and the meaning behind getting a tattoo were found to vary as much as the number of people getting tattoos. The similar thread running through the reasons for getting a tattoo, however, was that tattooing is a form of self-expression. Whether a person gets a tattoo 'just because he likes how it looks' or because it symbolizes something for them, the tattoo is a form of self-expression. The purpose of wearing this art on one's body rather than hanging it on a wall signifies a total commitment to what it stands for. It is the most permanent form of self-expression, with no escape from it. It is everywhere they go, they carry it with them, and it is a part of them. It is connected to one's mind and one's body for their time spent here on earth, and connected to their spirit, their inner essence forever.

In the first study, a questionnaire and statistical testing were used to evaluate the results. In the second, in-depth interviews and experiential research (where the researcher herself got a tattoo) were used to understand tattoo recipients' motivations.

Of course, there are overlaps among the methods in different studies. A researcher could use observational methods and a system of coding that allows for statistical analysis, or a questionnaire that asks for open-ended responses on the participant's feelings. It is important to emphasize that neither a quantitative nor a qualitative approach is best; instead, the methods used should be designed to fit the research question.

Likely the most familiar research approach for students is the scientific method, and this approach is often associated with a quantitative research perspective. Readers may question the application of the scientific method to qualitative studies or to the research in the social sciences in general, believing that the use of the scientific method is more firmly grounded in the natural or physical sciences. This would be a mistake: the scientific method is an approach to knowledge that is applicable to any research.

The scientific method replaced traditional or belief-based methods of obtaining knowledge (a reliance on religious doctrines, mythologies, or folk knowledge) and developed as a systematic approach to conducting research. A theory is accepted as true not because it is developed by an individual with political authority or a prestigious position, but because it has been shown to be true through observation and experimentation. (This does not mean, however, that the scientific method is infallible. Had the quantitative scientific researchers with Fisheries and Oceans Canada asked the right research questions, listened to the qualitative traditional ecological knowledge of inshore fishers, and not been prodded by the corporate profit-seeking imperative, they likely would have set much lower catch quotas and the Newfoundland cod fishery might not have collapsed in the early 1990s.)

The scientific method can be generalized into a series of steps:

1. The researcher notes some phenomenon worthy of study.
2. The researcher focuses this observation into a question.
3. The researcher crafts a hypothesis that limits the investigation of the phenomenon to an answerable question.
4. The researcher develops a research program, and predicts an outcome.
5. Research is conducted to prove or refute the hypothesis.
6. The researcher analyzes the results.
7. Conclusions are developed.
8. The results are published and disseminated to other scientists.

The research project seldom progresses smoothly through the eight steps: a researcher may set out a research question, and then refine it as more is learned on the topic. Data may be collected and interpreted, and then the researcher returns to collect more data. As a body of knowledge on a topic is developed, a researcher may seek to replicate the work of another scientist and fast-track the cycle to data analysis. Scientific research is ongoing, and the questions raised at the conclusion of a study often form the basis for new research.

The scientific method is a means of producing knowledge. Ultimately, the researcher seeks theories or laws on how the world works. While these laws may not be immutable (old theories can fall and new ones emerge), they represent the best abilities of the researcher to investigate and draw unbiased conclusions on an issue or phenomenon.

Why use a method, scientific or otherwise, for research? There are two primary benefits:

1. The use of a method adds rigour to a study. By working through a series of steps, instead of noting a phenomenon and then leaping to the excitement of a conclusion, the researcher has the opportunity of finding new information that may verify or refute the original contention. Through research design, analysis, and the presentation of findings, the researcher can take a systematic approach to finding cause–effect relationships and present this information in a manner that can be reconstructed by other researchers.

2. The scientific method attempts to remove researcher bias from influencing the outcome of the research. Ideally, the scientist will move through each step in the process with an increased sensitivity towards any foregone conclusions, and will have the ability to see where personal preconceptions are shaping the research. Research techniques such as double-blinding (for example, in drug testing, where neither the researcher nor the test subjects are aware of who is receiving the medication and who is receiving a placebo) are one means to reduce researcher bias. In the social sciences, a technique as simple as testing a survey on a focus group before mailing it to the entire sample set can reveal nuances in wording that bias the responses, and the researcher can correct for this partiality before conducting the study.

Ideally, the scientist is not so attached to the hypothesis that this will skew the methods used or the subsequent analysis: the hypothesis is a best guess at the results of the experiment, but unexpected results will still be acceptable and reported on without bias or manipulation. In the best constructed research, the scientists are able to separate themselves from favouring a predetermined result: the conclusions will remain true regardless of their personal leanings, and will be based on the completion of a rational experimental process and logical data analysis.

Finally, the use of the scientific method allows other researchers to repeat an experiment and determine if the results can be replicated or reproduced. Any variance from the expected results will require further investigation to determine the factors (experimenter error, misread data, typographic mistakes) causing the discrepancy. A theory may be proven false (a null hypothesis is an opposite hypothesis—it is a deliberate attempt by the researcher to prove a hypothesis false): this may not necessarily be a criticism of the research as many things once thought to be true by scientists (such as a flat earth) were later proved to be false.

The scientific method is sometimes criticized for assuming that the standard of impartiality is even possible—a skeptic would question how any study can be truly impartial, when the researcher is developing the hypothesis and setting out a research agenda to prove a belief held by the very same researcher. At best, the researcher may subconsciously design the study and interpret the data to fit the predetermined response; at worst, the researcher may manipulate the data to prove the hypothesis correct.

The question of bias is important: in social research—researchers may be unable to recognize or separate preconceptions that shape their understanding of both the research question and the results. Objectivity is sought in the research, but the difficulty of being truly value neutral must be recognized. Questions that should be asked in evaluating social research include:

*   Have questions on a survey been written in such a way that the researcher is more likely to obtain a desired response? Is the experiment designed to get only one result?
*   Is the selection of test subjects entirely random, or is it shaped to favour the response sought by the researcher?

Box 7.4 What Is True in Science?

One of the criticisms of research—particularly in the social sciences—is that 'truth' seems to be a fairly fluid concept, and research seems to be always disproving what was once accepted as laws or facts.

- We no longer believe in environmental determinism—that people are who they are based on their location on the planet—as a 'law', although this theory was popular in the early part of the twentieth century.
- Most psychologists today do not accept phrenology (reading the bumps on a person's skull) as a means of determining personality traits, although medical researchers in the early nineteenth century devised detailed charts relating parts of the skull and brain to particular skills, feelings, disorders, and the like.
- Most mainstream geologists would agree that the earth is not hollow, though they could have believed so in the mid-nineteenth century.

Simply because new evidence leads to new theories or knowledge, this does not mean that the previous research was deliberately wrong: it may be that more accurate testing brought different results, or the investigation of previously ignored data altered the findings. Science evolves with new knowledge and experimental evidence.

- Are the research methods transparent? Could anyone replicate the experiment and get the same results?
- Do the results make sense? The use of statistical analysis as a research technique does not equate with taking on a quantitative perspective or employing the scientific method. Poorly constructed data collection methods cannot be repaired by running chi square analyses or completing correlation testing.
- Are all the results reported—even those that do not fit the expected response?
- What conclusions are drawn from the research? Do these make sense, given the evidence?

Biased research does occur. Sometimes, researchers are oblivious to their own perspectives; at other times, it may be that the manipulation of results is intentional. As a student, professional, and scientist, it is important always to be a critical reader and observer—if the progression from a hypothesis to results seems to be missing a few steps or the links appear tenuous, be wary of the conclusions.

Chapter Review

This chapter examined the basics of research—who does research; what the key components of research are; what constitutes good research that reaches beyond simply asking a question to following a method and finding results. The key point in this chapter is to be cognizant of researcher bias: in the shaping of the hypothesis, the selection of research methods, and the analysis of data, good researchers will always be aware of personal bias and the ways in which their perceptions may be influencing the research. The ability to step away from one's personal biases is a good quality for a researcher.

Review Questions and Activities

1. Why is it important to develop a null hypothesis? How does this contribute to the research?

2. What is the difference between an inductive and deductive approach to research?

3. Which approach—inductive or deductive—is more characteristic of the application of the scientific method?

4. List four sources of knowledge. Think of examples for each of these sources that have shaped your behaviour or actions today.

5. Access an on-line journal in your field of study. After a review of several past issues (reading the abstracts and research methods for each article), do you find that most of the studies are quantitative in nature, qualitative, or a mix?

Chapter Eight

Recording Research

After spending valuable time conducting research, it makes little sense to neglect the production of an accurate record of your observations. In the social sciences, there are three types of research reports:

1. *observational reports* based on data collected in the field on human subjects;
2. *experimental reports* developed from information collected in a controlled environment, such as a lab, a survey, or formal interviews;
3. *field reports*, where the site characteristics are the focus of the research.

For any of these research reports, there are two key components: format and content. The format should follow a regular pattern, making it easy for the reader to find necessary information quickly. Content is also important: a research report is a fact-based document, not an essay on feelings or the researcher's reminiscences. The report is intended to document an event, experiment, or site. It is founded on information, non-biased, and accurate to the fullest extent possible for the researcher.

Regardless of the type of research report, before you begin writing your report, consider if you have completed each of the following:

- Are you finished your research? This may seem to be an obvious question, but pre-writing the research results can lead to researcher bias if the researcher refuses to accept any results that do not fit preconceived notions.
- Are you prepared to complete a draft? If possible, write the first draft of the research report from start to finish in one sitting. This allows the researcher to take a more comprehensive and coherent approach to understanding the research findings.
- Have you consulted all appropriate sources? Are there key readings in your area of study related to your research that need to be reviewed before writing? Will a good understanding of these works contribute to your research?

Box 8.1 Observational Research

Research requires that the social science researcher spend time with the subjects. However, the research can be conducted either directly with the participants, as with covert participation, or indirectly, as with a mail-out survey. Issues with each type of research are outlined below.

In **observational research**, the subjects are generally unaware that they are being observed. The researcher may be visible to the subjects but appear benign (a person in a crowd) or may be hidden from the subjects (separated by a blind or mirrored glass).

This type of research can be very time-consuming. The researcher does not control the research situation, and must wait for the human subjects first to be present at the site and then to behave in a manner that relates to the research. This may require many site visits until a body of research can be developed.

In **participatory research**, the researcher declares what he or she is doing with/to the subjects and interacts with them. The major issue in participatory research is that the subjects may change their behaviour to actions they believe are more acceptable to the researcher (both consciously and subconsciously), or they may even try to alter the research (again, either intentionally or not) by behaving in a manner that is entirely contrary to their normal behaviour. It is difficult to know for certain if subjects in a participatory research setting are acting in a manner that is entirely true.

In **covert participation**, the subjects should be unaware of the researcher's role. This type of research can be extremely time-consuming to set up—the researcher must find a way to be accepted into the research group and convince the participants that she or he is 'one of them' and not a researcher. A second issue with covert participation is the publication of the results: the subjects may feel betrayed and perhaps angry if they find that the person they believed was their friend or colleague was, in fact, only interested in them from an experimental perspective. Consequently, after the field research is completed, it is best to inform those who have been observed and ask them to check over the research findings. In some cases, of course, because of age or other demographic differences between researcher and subjects, this may prove difficult.

In **experimental research**, the researcher may be delivering a survey or conducting an experiment, but the interaction is as a neutral scientist, not as a participant or associate. As with participatory research, the respondents may intentionally or unintentionally alter their true responses to fit the response they believe is expected by the researcher. For example, if the researcher is using a questionnaire as the research method and asking questions on a sensitive subject (perhaps involving ethics or personal beliefs), the respondents may bias their answers to make themselves seem more neutral, when in fact they carry strong opinions on the subject.

- Have you formulated a clear hypothesis? If your research is complete, it is late in the process to be re-examining your research question. However, the start of the writing process (and the completion of the research) is an appropriate time to consider if the results obtained actually respond to the research question.
- Are you clear on the requirements for the draft? What are the expectations of your client (or instructor)?
- Have you considered the addition of tables, graphics, maps, photographs, or other illustrative data?
- What would assist in the analysis of your data?
- Have you recorded all citations? Have you done so in an appropriate format?

The sections contained in the research report and the order of reporting may differ among courses or work settings. As with all assigned writing, be certain of the criteria important in your program of study or workplace before completing any assignment. Specifics related to observational, experimental, and field research reports are considered in the following sections.

Observational Reports

In many social science studies, your subjects are viewed in their natural environment. The researcher observes (either as a direct observer, participant observer, or covert participant) the behaviours and interactions of the test subjects, and records the details as seen by the researcher.

Ideally, the researcher will make observations and record information immediately to ensure that data are not lost or influenced by other results or information. If this is not possible (for example, if you are a covert participant and embedded in your subject group), then your observations should be recorded as soon as possible. Do not delay in recording this information, as you will retain less and less of your observations as time passes.

Good research reports record events or conversations as they occur, in unbiased (as much as possible), non-evaluative text. Write as much as possible, letting the observations flow from your senses to the paper. As you write, some of the text may advance from *direct observation* (three skateboarders, mid-teens, all with baggy pants) to *analysis* (the dress of skateboarders potentially assists in defining this subculture) to the *researcher's feelings* (those pants must be uncomfortable). All of these notes are a valid part of the research report and should be noted. It may be that the research later focuses on only one aspect of the research (such as the relationships between subgroups at a skateboard park) but the research report can indicate that other observations have been made.

In sections on direct observation and analysis, evaluative words should be avoided. Evaluative words may be positive or negative: the 'awesome skateboarder' or the 'poorly skilled jumper'. While these may be included in the section on the researcher's feelings, every attempt should be made to separate comments based on fact from those based on bias or researcher perceptions.

Practical tip. In many observational studies, the subjects are never consulted on their participation in the research. The research is conducted in a public place and the subjects are observed while in this space. Care must be taken by the researcher not to publish information (photographs, explicit details on physical appearance, or descriptions of identifying characteristics) that would lead directly to a subject, unless the researcher has consulted with that subject and obtained permission for the subject's participation in the research as well as a release form to allow the publication of identifying results.

The following provides an outline for information that an observational researcher would record while in the field:

- *Location of study.* Where are you? What are the civic address, legal description, street intersection, compass direction, GPS coordinates—any identifiers that will allow another researcher to replicate or analyze the results. How is this site different from/the same as other research sites (if multiple locations are being used)?
- *Time of study.* The time of day may have an impact on participant behaviour and activities at the location (for example, consider the activity that may occur in a park during daylight hours as opposed to late on a Friday night).
- *Duration of the research.* How long was the researcher in the field? Are some periods in the field brief and others of long duration? Does the researcher only stay until certain behaviours are observed?
- *Series data.* Are these among several observations conducted over a period of time? It may be that the researcher's observations change with additional research. The chronological order of study notes may be an important component in the analysis of the findings.
- *Environmental conditions.* The temperature and atmosphere of the research site are of particular importance for some research questions. For outdoor research, this would be the weather conditions at the time of the research, along with a note on the conditions over a relevant time period (for example, is the research on the use of park space being conducted on the day after a major storm event or other occurrence that would cause patrons not to use the park?). For indoor research, the temperature, air quality, and ambient noise may have an impact on the research subject's behaviour.
- *Miscellaneous data.* Factors such as statutory holidays or unrelated events (a World Cup soccer celebration on the next street) may impact the behaviour of participants and should be recorded.

Other important field information to note in the research includes:

1. *The site*
    - What kinds of buildings or structures are located in the area?
    - What is the condition of these buildings? Are they in good repair or derelict?
    - Are there graffiti, broken windows, or signs of vandalism?
    - Are there any civic buildings or structures (like statues or street art) visible from the site? Does there appear to be government or civic care invested in the space?
    - What kinds of surfaces do you find in the space? Are there green spaces and places that welcome relaxation? Is the space characterized by hard surfaces like concrete and glass? Are there places to sit or is it a space that people pass through?
2. *Using your senses*
    - What are the sights, sounds, and smells in this place? Are they pleasant or unpleasant?
    - How do you feel as you observe this space?
    - How do you think others feel in this space?
3. *Other people/activities*
    - Who is in this space?
    - What kinds of people are characteristic of the space and appear to belong there?
    - What kinds of people would not belong in the space?
    - What are the primary activities of people in this space?
    - What else could you do/not do in this space?
    - Do the activities seem to be divided in any way (for example, are there areas that are used for only one kind of activity?)
4. *You as the researcher*
    - Do you think that anyone has noticed you (as a researcher)?
    - What impact do you have on the space?
    - What perceptions/values/biases do you note for yourself in your reflections on this space?

Once you have collected your data, the information can be drafted into a research report. The following provides a template that can be used for presenting observational data. Some sections (e.g., the context section that refers to previous research reports or recommendations for additional investigation, such as archival research) may not be required for your particular assignment. The template below is intended to be a comprehensive listing, and the researcher will choose those sections that are most relevant or required for his or her research report.

Observational Research Report

The observational research report will include some, if not all, of the following sections: purpose, context, background, data summary, methods, researcher's perspective, analysis, conclusions, and references.

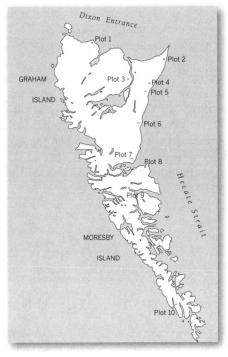

Figure 8.1 Graham and Moresby Islands
Research locations shown on the map (each plot point would be described in the text or in an appendix).

1. **Purpose.** This should be a short statement that describes the point of your research. What phenomena are you examining? Why? What is the hypothesis or research question?

2. **Context.** If relevant, previous research, academic studies, or published reports that are important to your research are noted. That is, are you conducting your study because it relates to previous research? Is there a context that the reader should be aware of?

3. **Background.** Beyond previous research, are there other forms of information available on your research question? Consider the following:

   • *Statistics.* Statistics Canada provides on-line demographic information and also publishes special reports that may be useful to your research. Local governments and provincial agencies may have important data related to your research topic that are readily available.

   • *Maps.* If relevant, mapping can add clarity to the research. It is easier for the reader to understand the location and relationship of places on a map than to have these described in the text. Beyond simple street maps, many local governments and provincial agencies have reference mapping available that provides information on zoning, environmentally sensitive areas, topography, soil quality, and much more. For example, Figure 8.1 shows the location of the research points on Graham and Moresby Islands. Marking the researcher's location on the map helps the reader to visualize the research. Information may also be available on thematic maps that illustrate issues such as levels of crime or income distribution that can add to the depth of your study. You may also find aerial photographs that present virtually up-to-the-minute pictures of the site.[1]

- *Bylaws or legislation.* What regulations are relevant to the study? With the Central Park study, would there be any regulations in the park that would influence the results that could be obtained?

- *On-file or archival information.* What information can be obtained from archival sources on the research question? Do any agencies have information on file that would be relevant to the research? Is such information freely available, or will the researcher have to consider filing an Access to Information request to view the data or documents?

- *On-line sources.* After investigating on-line academic sources (looking for previous studies or research), non-academic information may be found useful. While unpublished information may not have been vetted through a peer review process, there may be information useful to your research. Consider these sources from a critical viewpoint.

4. **Data summary.** This section records the details of your research as collected in the field. Generally, a summary is included in the text and an appendix provides the detailed information.

5. **Methods.** How did you approach and conduct your research? Provide a clear description of the methods used.

6. **Researcher's perspective.** Ideally, you have recorded all your observations as a neutral observer. However, it is important as a researcher to consider if your perspectives or biases have influenced any aspect of the research—the choice of locations, the hypothesis, the methods used, or the recording of data. If you are conducting research in an outdoor location and are cold and miserable, are your observational skills at their finest? If you are in a situation that you perceive as fearful or uncomfortable, you may see every action taken by human subjects as threatening. Note down your perspective, even if you believe you are conducting purely non-biased research.

7. **Analysis.** What are the connections among your observations? Are there comparisons to other situations or previous research? Note that this is not a section for the researcher's opinion: instead, this section seeks an analytical review of the information collected.

8. **Conclusions.** Restate any critical information or important connections. Conclusions should be based in the research: what was learned from the research, and how does this compare to what was known from the background materials? The research report should not be a vehicle to present the unsubstantiated beliefs of the researcher, but instead provides a technical, factual, accurate description of the research or situation.

9. **References.** List all references considered.

While observational research tends to be qualitative in nature, the rigour by which research reports are completed is no less comprehensive than for quantitative experimental reports. The methods of data collection must be clearly presented and should be replicable by other researchers. The hypothesis must be clear and precise, and the conclusions need to be directly related to the research and relevant background information. The old idea that the social sciences are somehow less 'scientific' than the natural sciences is best responded to by the production of transparent, rigorous, and detailed studies that meet the standards expected in high-quality research.

Experimental Reports

Experimental reports follow the same structure as observational reports; the key differences are in the methods used to obtain information.

First, an important initial consideration for the report is the audience that will read it. Is the report intended for publication? If it is, it is critically important to format the report in the manner required by the text or journal. If it is to be read by an audience of learned professionals, the references and details will be different than if it is to be read by a group of people without background knowledge on the topic. Be certain about the audience and write to respond to their information needs.

An experimental report presents the findings of a research study, and a critically important part of the study is the presentation of the methods used. Methods may include a survey, a statistical analysis of obtained data, or a report on the results of an experiment conducted in a controlled research setting. The researcher is providing information on the rigour of the study and the means by which the results could be replicated by another researcher. The researcher attempts to separate her or his opinions from the research and to present information perceived as being uninfluenced and unbiased.

The sections of an experimental research report are as follows:

1. **Hypothesis/problem statement.** The opening paragraphs clearly state the research question. The rationale for investigating the research question (why the researcher feels it is important) may also be presented.
2. **Background.** The researcher may expand on the reasons why the question is being investigated and explain the rationale for selecting the factors being studied instead of focusing on other potential causes.
3. **Literature Review.** It is important to reference key studies that have been conducted in the area of research, written in a narrative flow (not as an annotated bibliography). This section shows that the researcher is aware of previous research and is building on the findings of these studies.
4. **Methods.** In this section, the researcher will discuss:
   - The methods used to select subjects or participants in the study. For example, if the researcher has conducted a scientifically random sample of residents in a city, there will be a need to describe how the sample was selected and verify that the sample size is correct and representative of the larger population. Any issues in sampling will be presented (if the researcher feels that the sample could be questioned or if there may be legitimate issues of representativeness within the sample).
   - All tools for collecting information (survey questions, how interviews were conducted, and the set-up of the lab-based research).

Box 8.2 Famous Social Science Experiments

A number of experiments in the social sciences have achieved the level of 'classics', sometimes due to the extreme nature of the experiments and sometimes for their pioneering approach to research. Students are encouraged to consult the sources listed for more in-depth information on these classic studies.

The Milgram Experiment. Yale University psychologist Stanley Milgram describes his memorable study on obedience in his 1974 text *Obedience to Authority: An Experimental View* (New York: HarperCollins). His study was intended to measure the willingness of a subject to obey an authority (the experimenter) and the extent to which the subject would go before ceasing to accept the authority's orders.

In the experiment, two subjects were chosen to participate. Unknown to one of the subjects, the other was in fact working with the experimenter and was only playing the part of a subject. The experimenter pretended to randomly cast the subjects—one as the 'teacher' and one as the 'learner' (with the actor assigned as the learner). The learner was then strapped to a device that was said to give an electric shock—for every wrong answer to a series of memory tests, the teacher had to give an electric shock to the learner.

The learner proved not to be a good student, and the experimenter informed the teacher that the voltage was to be raised 15 volts after every mistake—although, in fact, no electric shocks were actually given to the learner; the learner only acted as though he was being shocked. The experiment escalated, with the learner requesting that he be allowed to quit the experiment, and the experimenter indicating that the subject signed up for the testing and is required to continue. As the experiment progressed, the learner/actor feigned increasing pain, begging for the research to stop. If the teacher indicated discomfort with proceeding, the experimenter used various verbal 'prods' to gain compliance, ranging from 'please continue' to 'you have no choice' and that the experiment would be ruined by non-compliance.

Among other findings, Milgram recorded that about two-thirds of the teachers would deliver electric shocks to the maximum (450 volts) that the equipment supposedly would allow. He also found that variations in the experiment (having the experimenter in the same room as the teacher) would cause increasing compliance among the subjects.

While universities would find it difficult, due to ethical considerations, to approve a study like this today, Milgram's experiment provided quantifiable data on the extent to which people will obey authority. It stands as one of the landmark studies in understanding human behaviour.

Zimbardo's Stanford Prison Experiment. This is another classic study of human responses, in this instance regarding incarceration. In the 1971 study, subjects were recruited from a newspaper advertisement seeking young males (to be paid $15/day to participate in a two-week experiment). The experiment was conducted out of the basement of the Psychology Department at Stanford University, which had been modified to resemble a jail.

The participants were interviewed and thought to be stable, 'normal' individuals, representative of the general population of white males. The subjects were divided into two groups—guards and prisoners. Both groups were given uniforms—the guards wore military-style shirts and mirrored glasses, while the prisoners wore poorly fitting hospital-style gowns and stocking caps (to replicate shaved heads). The prisoners were assigned numbers and were only referred to by these numbers.

The guards were given little information on how to run the prison, and were to use their best judgement. The prisoners, as well, were given little instruction, and were to behave as required by the guards. The experiment quickly escalated out of control. The guards developed inhumane methods for controlling the prisoners, and the entire experiment became increasingly humiliating and sadistic. The experiment was stopped after only six days because of fears for the prisoners.

Again, ethical approval would be difficult to obtain for an experiment in this format; however, the experiment stands as a classic in illustrating how 'real' an experimental situation can seem to subjects immersed in the experiment.

For more information, see Phillip Zimbardo's website: <www.lucifereffect. com>.

Snow's Cholera Experiment. In 1854, John Snow was a medical doctor living in London. Cholera was particularly virulent, and the medical profession blamed the cholera outbreaks on mists or 'miasmas' that carried the disease. Today, it is well established that cholera is spread in contaminated drinking water, but this was unknown in the mid-nineteenth century.

Snow observed that the disease appeared to occur in clusters. He plotted the occurrences of the disease, and observed that the tightest distribution was around a pump located on Broad Street in the Soho district of London. After obtaining a sample of the water, Snow found that it contained an unidentified bacteria, which he hypothesized could be the cause of cholera. He removed the handle from the Broad Street pump, and the spread of the disease was instantly halted.

This early study pioneers the use of plotted information; although done by pencil and paper, it represents one of the first applications of data to mapping—a precursor of modern GIS.

- An overview of how the research was done (how the experiments were conducted, how much time was spent on average doing the telephone interviews, how much time was spent by the researcher in the field).
- The sequence of events, as the way the data were obtained could influence the results (that is, the researcher may have been influenced by results obtained early on and this may have shaped subsequent research).
- How the results were analyzed and any statistical methods used to extrapolate beyond the obvious results (that is, the researcher may run cross-tab analyses that show the percentage of survey respondents who answered one question is dependent on their response to another question).

- Pilot testing or focus groups that were employed to give a baseline to the study.
- The methods used to obtain all collected data, and the researcher will identify any issues that may have had an influence on the results.
- The data in summary, with the details provided in an appendix.
- Tables and figures, if they add value to the presentation of the data.
- The reliability of the data. Does the researcher believe the information was accurately collected and that the data are valid? Does the researcher believe the information accurately represents the sample group and larger population?

5. **Results.** The results are presented, tracing the connections back to the research. This section proves why the research was valid and why the results make sense.
6. **Conclusion.** The often stated conclusion is that 'further study is required'. In the conclusion, the researcher will note any outstanding questions with the research, and will suggest areas for additional study.
7. **Bibliography.** All works cited or considered in the research are included. This does not include only works that are directly referenced, but also those that the researcher may have read and that may have had an influence on the researcher's ideas and perceptions.

The purpose of experimental reports is to present the data in a scientifically defensible manner, meeting the rigour expected of a scientific researcher. The format used and researcher precision in obtaining and reporting on results is critically important in justifying the logic and value of the work.

The Internet is increasingly being used in the social sciences to obtain survey results or to conduct experiments. For example, the Psychology Department at Hanover College in Indiana provides web-based access to a wide range of surveys that can be completed on the web. Researchers can forward their study to the department, and if it meets the college's standards it may be included on the web portal. This allows for access to a wide array of survey respondents or participants: the experimenter can identify the characteristics of the required subject (by socio-demographic characteristics, geographic location, or any other parameter established by the researcher), and wide familiarity of the website is anticipated to increase the likelihood of reaching these subjects.[2]

Of course, the critical issue with web-based research is the inability to verify any of the information provided by the respondents. The researcher must have a level of confidence that the information provided is accurate, and that no attempt is being made to intentionally falsify the results provided. As well, the respondents are limited to individuals with web access and a predilection for responding to surveys or participating in experiments—neither of which may be characteristic of the larger population.

Given the likelihood of increased use of web-based research in the social sciences as a means of accessing subjects, the methods used and the care taken in analysis will be critically important.

Box 8.3 On-line Research

The Hanover College website contains a wide array of interesting surveys and experiments that web viewers may participate in. Recent research includes:

- a survey on superstitious beliefs by Bethany T. Heywood and Jesse M. Bering, Queen's University, Belfast;
- a video test, where respondents are shown a brief video, then tested on their recollections by Joshua Hartshorne and Yuhong Jiang, Harvard University;
- a study on rating faces, where participants are asked to rate their reactions from 1 to 7 on four different scales (intelligence, physical attractiveness, trustworthiness, and distinctiveness) by Brian R. Spisak and Hannie van Hooff, University of Kent;
- a study on thinking and reasoning by Erin Beatty and Valerie Thompson, University of Saskatchewan.

Field Reports

Field reports, simply, are documentation of your research experience conducted in the subject's environment instead of an office, lab, or library. In a field report, the environment is a critical part of the study. Your research may involve human subjects as it did for an observational report, but the focus of the field report is on understanding the environment surrounding the human subjects. These types of field reports are most akin to site research in the natural sciences—just as a biologist might investigate a plant or animal species, a good field researcher in the social sciences is seeking to collect live data on the research area.

A field report is a means of recording in detail all the information available on a particular site or situation as experienced by the researcher. If you are conducting field research you will seek to experience the event or place as a scientist. Every detail becomes important and is recorded in your report, and each is analyzed for meaning. How does an aspect relate to other aspects noted? What does a behaviour or site characteristic mean? Where else has this been observed? The primary purpose of a field report is to record information on the site or situation with such accuracy and in such detail that others will be able to understand the site from your well-written field report.

As a field researcher, your observation skills are working at their maximum capacity. While many field reports contain only a visual survey of the situation or site, better reports consider the site more experientially. What sounds do you here? Are any odours observed (e.g., is the site near a landfill that is hidden by tree cover)? Tactile and taste experiences may be more difficult to seek out, but should be considered.

In addition, the field report must examine the site or situation in four dimensions—length, width, height, and time. In a site survey, consider the impact of off-site situations—an industrial site many kilometres upstream could have impacts on the creek traversing your site. Consider the land uses on adjacent sites, both seen and unseen. Examine the tree cover from the ground level to the tree tops. Consider the time frame when your research is conducted—would the results be different if you visited the site at a different time of day? A different time of year? Measure and accurately record distances, both from the site boundaries and among features on the site.[3]

Field reports are written with an end in mind. Generally, the researcher is completing an assigned task or conducting original research directed towards a purpose. This purpose will shape the observations of the researcher. Therefore, your field report must contain an analysis of any bias or preconceived notions being brought to the site or situation. Carefully examine if your expectations are shaping your experience on the site. As a scientist, do your best to attempt an unbiased approach to the research and experience the site or situation from a neutral position.

The following provides a template for a field report. As your experience in the field increases, you will likely modify this template to suit both the subject or site and your style of research. In addition, you may modify the report based on the requirements of your study.

Practical tip. Unless you are conducting field research on a site owned by you or your client, you will need the permission of the land or building owner to access the property. While free access is permitted to most public spaces, there may be bylaws or legislation in place that will limit your research. In addition, while some spaces appear public, there may be restrictions on use:

- downtown streets with 'no loitering signs';
- parks with signed hours of operation;
- public buildings that do not allow access without an appointment;
- pseudo-public spaces such as malls or common areas outside of commercial or residential complexes: while these spaces appear to be public, they are not, and your use of the space is restricted;
- gathering places, such as coffee shops and pubs: the owner may not support your use of the site for research;
- gang territories: access to the territory may require the permission of gang leaders;
- vacant lots: while a lot may appear unused, it is private land.

Information on landownership may be obtained either from a contact on site or at a land title office. Accessing a privately owned site without permission is trespassing.

Conducting Field Research

The primary advantage of field research is the proximity to the issue under investigation. As a researcher, you have an opportunity to observe a phenomenon in its natural location. This advantage is also the dilemma of field research, as your presence in the field may have an impact on the research subjects. Particularly in research involving human subjects, your methods may have a profound effect on the results obtained. Even if your role is that of an observer only, your presence may impact the subjects' behaviours or interactions, especially if the site of your observation is not a place you have normally frequented in the past. A more participatory role, where the subject is aware of the research, will definitely have an impact. Even as a covert participant, your actions and behaviours will shape the reactions of the subjects of your research. In addition, the more involved the researcher becomes with the human subjects, the more difficult it is to avoid researcher bias. While field research can be rewarding, it can also have emotional and intellectual impact on the researcher, can be wildly time-consuming, and can also be dangerous (see Box 7.3).

In the social sciences, your fieldwork will result in either of two types of field reports: those involving observations of and data collected from human subjects, and those that provide an analysis of a defined geographic area or phenomenon.

Site Research

Site research begins with the premise that a site has qualities that can be assessed and understood by the researcher. The field report documents the site's qualities and characteristics and applies this learning either to document current uses or to propose future uses.

Data that should be collected and reviewed before the site visit include any available aerial mapping, topographic maps, soil surveys, and infrastructure maps that illustrate roadways, access points, utility corridors, easements, and well locations. If available, historic maps can be researched to investigate change on the site over time. In addition, a property description map that shows buildings and structures, property lines, dimensions, and adjoining uses will be valuable for orienting the researcher to features on the site.

Bylaws and legislation will also impact permitted uses of the site. Local government planning departments can provide information on current zoning and land-use designations, and provincial agencies will have information on other regulations that control land use, such as watercourse protection legislation. The researcher should investigate not just the regulations specific to the parcel, but to surrounding sites as well. For example, a neighbouring site may be vacant and well-treed and provide significant natural amenity value to the subject property. However, with investigation, the researcher would find that the neighbouring site is zoned industrial and the local government is currently considering an application for a noisy and noxious rendering plant. Therefore, the researcher should investigate not just the current regulations but also inquire as to whether any changes are anticipated by the government agencies. The researcher will obtain valuable information by interviewing senior researchers, government officials, or local residents with knowledge of the site and area. The information collected can then be triangulated to evaluate the accuracy of information sources against each other.

A good field investigator will report on much more than the obvious physical characteristics of the site. Less tangible characteristics are also important, and the site report must consider all the natural and human-caused characteristics of the site, including:

- *Watercourses and water bodies*: creeks, streams, wetlands, storm water drainage, ponds, lakes.
- *Land features*: peat bogs, rocky outcrops.
- *Environmental features*: eagle nests, heron roost trees, diversity of plant species, evidence of animal habitat.
- *Topography*: steep slopes, variation in slope across the site, areas of erosion or subsidence.
- *Orientation*: the site's orientation to the sun and compass directions.
- *Views*: long-range and water views, desirable or undesirable vistas.
- *Elevation and climate*: the on-site climate characteristics can vary widely with elevation and micro climates on the site should be documented (areas under permanent shade, west-facing sunlit hills).
- *Buildings or structures*: both current structures and any remnants of human activities (old building foundations, abandoned well heads, derelict power poles).
- *Signs of use*: other signs of human occupation or use, such as the presence of non-indigenous plant species, culturally modified trees, abandoned access roads, walking paths, or refuse.
- *Off-site characteristics*: While the site under investigation may be defined by legal property lines, the collection of data will consider factors beyond these boundaries. Are there proximate uses that will impact the use and enjoyment of the site under investigation? For example, the sound from a gravel crushing operation located two kilometres away may be heard from the site; a nearby aged landfill may be leaking contaminants into the soils below the property; the odours from an intensive feedlot operation some distance away may be detectable from the property if the property is downwind.
- *Context*: Where is the site in relation to other uses? Is it proximate to urban recreational and shopping amenities? Are rural amenities (such as peace and quiet) characteristic of the site? The researcher will examine the relationship of the site to the larger area for factors that could add or detract from the use of the subject site.

As with subject research, the purpose of the site report is to collect accurate, unbiased information that will be used either to document the site or to provide recommendations on a proposed use for the site.

Box 8.4 Checklist of On-site and Off-site Characteristics

✓ **On-site Characteristics**

- Auto access and parking
- Topography: level? hazard lands?
- Utility lines, poles, sewer connections, water and gas mains
- Property lines
- Watercourses
- Buildable areas
- Drainage patterns
- Sources of contamination
- Evidence of human habitation
- Evidence of underground features (mine shaft)
- Utility trenches, well heads
- Areas of fill materials
- Materials on site: gravel, rock, etc.
- Agricultural values?
- Aesthetics: views of desirable or undesirable features
- On-site uses: entrenched public use of vacant parcels (pathways, bike jumps)
- Site diversity: vegetative and wildlife inventory
- Individuality: dominance of a landscape element
- Environmental features: eagle trees, heron roosts, evidence of habitat
- Cultural features: built and non-built environment that is visually attractive/unattractive, has impact/no impact
- Densities: what density is proposed for site?
- Scale: how will the proposed site development relate to surrounding development?
- Contrast or connectedness: how will the development fit with surrounding development?

Off-site Conditions

- Adjacent land uses: impact of your development on these sites or impact of adjacent land uses on your site
- Noise, odours, effluent
- Drainage patterns

(Continued . . .)

- Views onto the site from other properties
- Impact of natural features (e.g., in rainshed of mountains, located in a floodplain)
- Proximity of amenities
- Proximity of other buildings or structures
- Proximity of community services (are water and sewer services available?)
- Proximity of utilities (is there power to the site? cable?)
- Impact of ecological features located on adjacent lands (eagle trees, habitat)
- Watercourses on adjacent lands
- Distances for travel
- Vacant parcels: do not assume that a vacant parcel is parkland
- Major plans known for the area: highway realignments, subdivisions, industrial uses, expansion of a commercial development
- Zoning amendments: changes in uses being proposed or under review

Use of Technology in Research

Photographs can be an invaluable tool in capturing the behaviours of human subjects, sharing site details with other researchers, or reminding the researcher of site characteristics. Photos can also assist in showing the relationship of objects to one another, and the inclusion of a measurement device in the photo can provide scale of the photo.

However, not all characteristics can be accurately recorded in photographs. A photograph is a two-dimensional rendering of a three-dimensional subject, so detail and context are immediately lost. In addition, a photo cannot provide other information that may be critically important to the research: music blasting from an adjacent apartment, the smells of the rendering plant over the hill, or the crunch of dry grass under the researcher's boots. In some cases, video is a better option as it allows for the addition of sound and provides a better 'remembering' of the researcher's experiences on the site. However, neither a photo nor a video can always accurately capture the nuances of human interactions. Photos and videos should be considered valuable tools in field research but not the only means of recording data.

The ability to sketch a situation or relationship can be useful to the researcher, particularly once back in the office or lab. Minimal drawing skills are required— what is important is to produce a drawing that will assist in your analysis.

Other devices, such as GPS units, sound recorders, surveying equipment, binoculars, and sampling equipment are certainly useful in the right context. However, the observational skills of the researcher remain the most important tools carried into the field. Trust your instincts and use all your senses!

Chapter Review

This chapter has considered three types of reports, developed around three common research methods in the social sciences. The first, observational reports, asks that the researcher expand his or her understanding of the research site and human subjects to factors beyond the obvious. The second, experimental reports, speaks to the rigour required from scientific reports intended for publication. The third, field reports, sets out a checklist and parameters for researchers doing work where an understanding of the site and setting is critical to the conclusions drawn in the research.

In any report, knowing the intended audience is important. Write the report to meet the requirements and needs of the readers.

Review Questions and Activities

1. What are the key differences between observational and experimental research reports? Between observational and field reports?

2. What is the primary concern with doing research that involves on-line responses to a questionnaire? What can be done about it?

3. Why is it important to write a report for the intended audience?

4. Pick a setting where you frequently spend time—the bus stop, cafeteria, or student meeting space. Over a one-week period, spend time in this space as an engaged observer. What do you notice that you have never noticed before? Do your observations change as you conduct more research?

5. How can a social scientist know the attitudes of participants in observational research?

PART III

Presentation Skills

In the following chapters we consider public speaking from a holistic perspective, considering speech and voice production, non-verbal communication, the use of presentation space, and finding your presentation style, and identify ways not just to survive public speaking, but to succeed to new levels of accomplishment and professionalism.

Chapter Nine

The Fear of Public Speaking

For most of us, conversation is a normal activity. We chat with friends, participate in class discussions, order take out, and talk on the phone. Most of us (some more than others) speak every day and with little anxiety to individuals and groups. We competently form words and sentences, impart meaning, answer questions, and complete thoughts with little effort. However, when asked to 'speak publicly'—that is, 'to orate' in front of a group of people in a more formal environment, our reaction can range from slight nervousness to a full-blown anxiety attack. In fact, researchers contend that public speaking is the most common social fear in North America—most North Americans would rate themselves as having a negative reaction to public speaking.[1]

Public speaking evokes such a strong reaction for two reasons:

1. We equate public speaking with conversational speaking, and expect a speech to flow as easily as a discussion between two colleagues. The differences between conversational speech and public speaking are not recognized or developed; therefore, when placed in a public speaking situation the speaker is unable to perform to expectations.
2. Or the opposite: we recognize the differences between public speaking and conversational speaking, that public speaking requires skill development and practice, and we know that we are not prepared.

Instead of practising public speaking and working to develop core skills, many put great effort into practising the avoidance of it. While evasion can be a useful short-term solution, dodging opportunities to speak will place limits on careers and life experiences. The ability to stand up and speak in front of a group of people—whether a handful of colleagues in a classroom or hundreds of people in an auditorium—is a skill that can be learned and mastered. While it is true that natural performers may operate at higher initial skill levels, and it is also true that public speaking may not be everyone's favourite activity, anyone can learn to be an effective public speaker (i.e., one who is understood by the audience) through practice and technique development. Understanding the mechanisms of speech (how speech is produced) and of speeches (what to say and how to say it) are the two key components of becoming a competent public speaker.

Your reaction to a public-speaking assignment and your performance are under *your* control—does the presentation of a public-speaking opportunity cause excitement or dread? If your reaction is less than positive, would you like to change it?

The Fear Response

Psychologists believe that humans are born with two innate fears: a fear of falling and a fear of surprises perceived as harmful, such as loud noises, rapid movements, or sudden darkness.[2] Amusement parks are built on these fears: roller coasters mimic falling, the success of the haunted house is measured in how many times you are frightened by ghouls popping out when you least expect it. These fears are innate because they are useful: we can get hurt by falling so we naturally protect ourselves from it, and we have heightened reactions to unforeseen events to ensure our personal safety. These would appear to be reasonable fears, ones that have allowed us to persist as a species.

Humans share other common fears: across cultures, people (quite rationally) tend to fear those things that may kill them—venomous snakes, spiders, and strangers. People tend to hold all of these qualms regardless of their location on the planet. It is likely that these fears develop both from experiences, where we see first-hand the problems caused by snakes, spiders, and strangers, and from beliefs, where we are told to be afraid by a person in authority and we acknowledge the statement as truth.

If we accept that our only instinctive fears are the above noted two: falling and surprises, and if we accept that other common fears focus on things that can kill or harm us, the question arises as to why so many people have a fear of public speaking. It is not innate, and we will not die from speaking (unless we evoke an extremely strong stress response). Therefore, the fear of public speaking must be either consciously or subconsciously developed.

It may be that an initial poor experience led the speaker to develop a fear of speaking, or it may be that the speaker adopts a fear based on the experiences of others. In either case, the fear of public speaking is a learned behaviour, one that can be unlearned through skill development and practice. Just as you might learn to lessen a fear of spiders by systematically desensitizing yourself to them (first drawing a picture of a spider, then viewing photos, then standing near a spider, and culminating with holding a spider in your hand), a speaker can desensitize a fear of public speaking by working through helpful exercises and experiences.

Although a fear of public speaking is an assumed condition, the reaction of people to public speaking is no less real than a reaction to any other fear stimulus, and a debilitating level of fear of public speaking has warranted its own label: *glossophobia*, from the Greek words for 'tongue' and 'fear'. Our reaction to public speaking is the same as our reaction to coming face to face with a taipan, a highly venomous snake, in the Australian outback:

- The body tenses.
- The eyes dilate to take in more information.
- We become more alert to movement around us.

- Reactions can become extremely fast or slow, depending on the person.
- The body releases approximately 30 different hormones intended to deal with the perceived threat.
- Blood pressure and heart rate rise.
- Visible signs of stress may be evident, such as shaking in limbs or voice or facial flushing.

This response, called the 'fight or flight' effect, is a useful one if an individual is in danger. Reactions are immediate and initiated without conscious thought, and the individual becomes prepared for defence against the threat or for exit from the situation. However, this response is not so useful in public speaking. While energy is required to be an effective public speaker, misplaced energy (focused on fighting or fleeing) detracts from the speaker's ability to concentrate on the presentation.

It may even be that the speaker has this reaction well before the public-speaking opportunity. Just the idea of public speaking can trigger a full-blown stress reaction. The speaker may predict all sorts of negative events: losing notes, freezing at the lectern, non-functioning audio systems, or unwelcome reactions from the audience. The speaker may then seek ways to avoid the public-speaking opportunity or may proceed with such powerful beliefs that the imagined negative situations either are created consciously or subconsciously. For example, a speaker may be so unfocused that she is unable to remember to flip a switch to power a PA system, thereby realizing her fear of technological collapse. We have a powerful ability to shape our own behaviour and create outcomes: dwelling on negative outcomes is, unfortunately, the way to guarantee them.

The speaker may look for ways to avoid the event or even seek out chemical substances to limit his reaction. While avoidance behaviour may be useful when in a dark alley potentially full of shadowy strangers, it is difficult to completely remove oneself from public speaking in an academic environment or the workplace. At some point in your career, you <u>will</u> be required to present, and it is a career-limiting decision to avoid all opportunities. A better decision is to resolve that you will triumph over any learned fears of public speaking and develop strong skills in performance that will allow you to be an effective speaker.

What can we do to minimize a 'fight or flight' reaction to public speaking? We can learn techniques to deal with this reaction, and we can increase our confidence as public speakers through practice and skill development.

Fear Reactions

Put yourself into one of the following situations (select the one you would find the most stressful). Try to imagine your reaction to it as vividly as possible:

1. Stuck in a traffic jam on the way to an interview for your dream career, there is *no way* you are going to make it.
2. Realize that the exam you thought was next week is scheduled for today, right now. Get out a pencil.

Notice where stress is causing tension: are you tensing the muscles of your neck, shoulders, or arms? All over? For a short period of time, you reacted as if you were placed in a dangerous situation, and your body reacted as outlined above. Work to understand your own reaction to a stressful situation. For many people, tension is first felt in the jaw—we clench our teeth and neck muscles (it is likely that our ancient ancestors did so to be ready to attack, while a more likely modern-day reason is to protect expensive orthodontic work).

Now, consider how this fear reaction will impact your presentation abilities. Increased tension, fast or slowed reactions, body shaking, and face flushing are not actions that will contribute to your success in delivering a message to your intended audience. If the idea of presenting causes any of these stress reactions, techniques are available to consciously release this tension, and a speaker can practise these techniques to shift quickly back into a productive presentation state (see Boxes 9.1 and 9.2).

Calming Techniques

What is the most powerful technique for calming the speaker? Learn to breathe out. Humans hold their breath when they are feeling threatened. This was useful when one was hiding in the jungle from a sabre-toothed tiger, when any motion could

Box 9.1 Three Ways to Immediately Calm the Speaker

1. *Relax the tongue.* A simple exercise for immediately inducing calm: relax your tongue. While it sounds overly simplistic, when we are tense or having a stress reaction, we tend to push our tongue onto the roof of our mouth or press it hard against our teeth. This causes the speaker difficulty in the jaw, shoulders, and upper body. Think of a situation that you would find stressful, note your tension, and then consciously relax your tongue. You will find this immediately produces a greater sense of calm.
2. *Nostril breathing.* While waiting for the beginning of your presentation, sit with one hand along the side of your nose. Close one nostril with your index finger, and focus on your breathing through the other nostril. Two things are happening: you are now focused on your breathing and you have reduced your airflow. You will find that you start to take deeper, fuller breaths, which increase a feeling of calmness. Take 10 breaths, then switch sides. You can use this technique fairly surreptitiously if practised in advance.
3. *Muscle tensing.* A third covert relaxation technique is to tense all the muscles in your legs—this can be done sitting down (tense thigh muscles, calf muscles, ankles, and feet) or standing up (focus on tensing thigh and knee muscles) while you are waiting to be called for your presentation. Tense the muscles and hold for a count of 10, then release. The feeling of release will cause the rest of your body to relax, also. Keep repeating until you are presenting.

Box 9.2 Stop Knee Shaking

Knee shaking is a common response in some presenters, where the extra energy created by the presentation is revealed through knee or thigh shaking. If you are behind a lectern, press your knee into the lectern (be sure it is well fastened and will not shift). If you are standing without a lectern, use your muscles to pull your kneecaps as high as possible, hold for a count of 10, and then relax. This can be done while presenting and will not be visible to the audience (unless you have chosen to wear a fairly short skirt). Relaxing the muscles will cause the shaking to stop.

As with the techniques described in Box 9.1, it is a good idea to practise these exercises in advance to make them fully useful in a presentation.

mean death. It is not useful when you are standing in front of an audience ready to begin a presentation. When stressed, we tend either to hold our breath completely or to revert to very shallow rapid breathing that does not provide sufficient oxygen for thinking and certainly not enough to fuel a powerful and confident voice.

Without adequate breath, the speaker has a new issue to feel anxious about, the voice is quiet and often shaky, and the presentation comes across as timid or fearful. A self-perpetuating cycle begins, where the speaker stops breathing, the lack of oxygen causes physical stress, this stress is interpreted as a fear of public speaking, the speaker stops breathing, and so on.

Breathing is the foundation of relaxation and also the basis of voice production. To release tension, you need to breathe out. Try this exercise now:

Take a deep breath in. Hold your breath as hard as you can. Now, while you are holding your breath, say: 'I am holding my breath in as hard as I can' without letting any breath out. This is very hard to do, and you will note the strain in you voice and tension throughout your body.

Now, take a deep breath, and while you are breathing out, say in a normal conversational voice: 'Now I am not holding my breath and the difference is amazing!' You will note the feeling of calmness that comes from breathing out.

Become conscious of your breathing patterns. If you are feeling tense, recognize if you are holding your breath. Let it go and keep breathing. Concentrate on taking deep, full breaths that use your entire lung capacity.

The second most powerful technique for calming the speaker is *practice*. Virtually any task becomes easier with sustained and ongoing work at it. Psychologists use a technique called 'exposure' to deal with fear responses, where the subject is gradually brought in closer contact with the feared object, whereby the subject learns to refrain from immediately entering into a full stress reaction to the stimulus.

Consider using exposure techniques to improve presentation skills. Practise your speech in advance (at minimum, four to six times, running through the entire speech) until you feel that you have mastered the flow and content of your presentation. Listen to your voice and strive to keep the speech flowing. At first, you may feel uncomfortable practising your speech aloud, but it is important that you do because reading your text to yourself will not adequately represent how the speaking event will feel.

Once you have successfully presented to yourself, build on your experience. Take small steps to increase your range: speak in front of a mirror, then present to a friend or colleague. Public speaking will become more familiar and more manageable.

Cognitive psychologists also refer to 'fear extinction' where subjects learn to have a new response to a fear stimulus. That is, the subjects identify what they are fearful of and then relearn their reactions to those stimuli. For public speaking, if your initial response is the 'fight or flight' reaction, give yourself the opportunity to try a new reaction. How would you *like* to react to an opportunity to present your ideas to an audience? What would it feel like to share information on a topic you feel passionate about?

Gradually increase your exposure to public speaking. Joining Toastmasters or another public-speaking association is an excellent way to gain exposure in a safe environment. You may also seek out presenters with good presentation styles and talk to them about their skill and technique development. Invariably, you will find that a good presenter is someone who works at it.

Practical tip. Set your own performance targets: this week, speak out in class. Next week, find an opportunity to read aloud to someone. Speak to a stranger at a gathering, using a statement you have consciously prepared in advance (of course, working within the bounds of propriety and safety). After that, work on preparing short presentation statements that you consciously practise, improve upon, and then present at a public venue (perhaps speaking at a city council meeting or other event).

The fear of presenting can be manipulated through ongoing and consistent practice. Use the nervous energy created by the presentation opportunity to add life and power to your performance.

Relaxation and Tension

While it is desirable to be a calm speaker, some tension is required to bring energy and forcefulness into a presentation. Consider the placement of human activities across a spectrum ranging from full relaxation to full tension. A 'normal' state would be in the middle of the spectrum, representing the balance of tension and relaxation needed to get through regular 'alert' activities such as note-taking in the classroom, driving to campus, or performing non-routine work tasks. One cannot (or at least, should not) achieve full relaxation during these functions—a certain amount of tension is required to stay alert while performing routine tasks.

Box 9.3 Energy for Performance

Full Relaxation	Increasing Relaxation		Mid-point	Increasing Tension		Full Tension
			Energy for Public Speaking			
No Energy	Lowest	Lower	Normal	Higher	Highest	Over-energized
Unproductive State	Minimally Productive	Low Productive State	Normal Productivity	Performance State (Comfortable)	Performance State (Energized)	Unproductive State
• Sleeping • Meditating	• Passively awake • Watching non-engaging TV	• Performance of repetitive or routine tasks	• Energy needed for work, school, driving • Engaged in familiar activities • Normal alertness	• First level performance state—raised awareness • Higher energy, creativity increased • Thinking more quickly • Reactions faster	• Second level performance state—high awareness, brain is racing • High excitement • May take actions that are viewed as beyond normal range of activities	• Perceive situation as dangerous or life-threatening • Too much excitement

At the left of the spectrum, full relaxation would equate with sleep or a deep meditative state. At the other end, full tension is represented by a full-blown 'fight or flight' reaction with adrenal glands pumping, eyes fully dilated, heart rate increased, and muscles fully energized.

Neither extreme is the best state for performance, nor is the safe centre zone of the spectrum. The speaker should target a state to the right of centre—more tension than is required to complete normal activities, but not so much that the body's involuntary systems are thrown into a stress reaction. In these higher tension states, awareness, creativity, and resourcefulness increase and the speaker feels 'on'. The standard of performance is increased from an everyday conversation to a performance where the audience can feel the energy of the presenter.

Box 9.4 Presentation Self-assessment

At the completion of every presentation, take a few moments to critically assess your performance. How did you do? Be honest, but without malice, and respond to the following questions:

- What went well in the presentation? Give yourself credit for having been understood and for having spoken clearly.
- Where was I better than in my last presentation? Consider each presentation to be an opportunity to add to your skill set as a public speaker.
- What would I do differently next time? Would you allow more lead time at the venue to ensure that technical issues (like lighting and sound systems) work smoothly? Would you allow more or less time for questions from the audience? Consider actions that you could take to elevate your next public-speaking performance.

Keep a written record to allow you to track your progress as an effective public speaker and work on those aspects you would most like to change about your performance. And remember that public speaking is a skill that can be developed with the application of some effort and a bit of practice.

Increasing Confidence

Each opportunity to speak should be viewed as a chance to increase skills and become a more professional public speaker. One successful speech can elevate the confidence of the speaker, changing the speaker's perspective from negative to positive (or at least to less negative). Eight methods and techniques for improving the likelihood of a successful speech are outlined below.

1. *Be prepared.* Practise as much as you can well in advance of the presentation. With practice, you will develop ways to reduce the time needed to prepare for a presentation, but be sure to allow yourself sufficient time to complete research, create

materials, and develop speaking notes for the presentation, and for rehearsal. There is a difference between having some nervous energy about a presentation that you are ready for and knowing that you are insufficiently prepared.

The day before your presentation, run through all the materials needed and make sure they are ready to go: complete the photocopies of handouts, and save the presentation in more than one format (see 'Plan B', below).

Plan to be completely ready 24 hours in advance to leave yourself time for any unexpected complications. Organize your presentation in your briefcase or backpack: ensure that you have your speaking notes, presentation, hand-outs, and any other materials or tools needed (like research documents or a laser pointer). Do not leave anything to chance that will cause you anxiety as you move into your presentation.

2. *The 'Plan B' rule*. To increase your confidence, always develop a backup plan. What will you do if your jump drive will not work? Have your presentation burned on a CD as well. What will you do if the projector bulb burns out? Have key slides on overheads and work from an overhead projector, or be prepared to work without any technological assistance. What if your laser pointer quits? Have a backup, or ask the audience if anyone has one (it is surprising how many people carry one). Brainstorm a list of 'things that can go wrong' and what you will do if they happen.

3. *Control the environment*. Check the presentation space well before you are presenting (preferably, when the room is empty). Figure out how the pro-jector and lighting work and how the sound system operates. Be certain to understand how the electricity in the room works—what light switches do

Box 9.5 Technique to Control Emotions

You may find yourself in a situation when presenting where your emotions get the best of you. If you are tearing up or feeling your throat constrict from emotion, try this simple technique. Look above the heads of the audience at a point on the furthest wall, exactly in line with your eye line. Concentrate on this spot for five seconds, and then adjust your eyes to look slightly upward. Do not look so far upward that the whites of your eyes are showing to the audience (as this can be quite disturbing for audience members) but look up just enough so you can regain control of your emotions. Pull your shoulders slightly back and keep your chin parallel to the floor.

Try it now. Five seconds may feel like a long time, but it will not seem so to the audience. In fact, pausing briefly is generally interpreted by an audi-ence as control—the speaker appears calm and confident.

Our normal reaction in a highly emotional situation is to look down, bow our head, pull our shoulders up, and collapse inwardly. This creates an emo-tional cascade effect, where our brain recognizes that our body is physically folding and lets the emotions follow.

Do not allow yourself to look down or fold; instead, the act of looking slightly up stops your eyes from tearing and assists in regaining control.

you need on or off? Which electrical outlets work (sometimes, coloured plug casings indicate that an electrical switch also needs to be turned on to make the outlet 'live')? Is there a trick to using this particular projector (some need to be turned on before the laptop is hooked up, and some after)? Do you need an extension cord for your laptop?

These all are issues that you do not want to be dealing with while the audience is sitting patiently waiting for your presentation. Credibility with an audience is quickly lost if the speaker has technical difficulties with a projector or microphone.

Ideally, try to run through your presentation from start to finish in the classroom or boardroom where you will be presenting at least a day before your presentation. Doing so creates the psychological advantage of familiarity; if you are used to working in that space and know that you can complete a successful presentation you will do better before your audience. In addition, a rehearsal will highlight any problems in the room, such as a poor-quality sound system. You may be able to develop your Plan B (even to bringing in your own speakers) if you know about such issues in advance.

Practical tip. Be aware of situations outside the presentation space that may have a negative impact on your performance; for example, the noise from an adjoining classroom or odours from a cafeteria can be distracting for your audience. You may be fortunate enough to have technical support at the venue, but, ultimately, the audience looks to the speaker for a professional presentation.

4. *The rule of three.* No matter how informative a presentation is, most people in the audience will be able to retain no more than three key points. As the presenter, you have control over what the audience retains as these key messages. Note these three points at the beginning of your presentation, structure the presentation to work around the three points, and then summarize them at the end. Do so in such a way that the audience does not feel that the presentation is following a template or format; the repetition through the presentation should feel natural, not contrived (see Chapter 12 for techniques).

5. *Timing.* You do not have to cover *everything* in your presentation. Often, a novice speaker will try to cover too much information in too short a time frame. If you are doing a 10-minute presentation, you do not need 30 minutes of material.

    How can you know if your presentation will fit the required time frame? Rehearse it. This does not mean reading your presentation to yourself while quietly at your desk. Most people read faster than they speak, so reading (particularly silently) will not give an accurate estimate of the length of the presentation. Without rehearsing, many speakers will realize too late that they cannot complete the presentation, and will spend the last two minutes trying to race through the remaining key points. The speaker then becomes flustered, the presentation loses focus, and the audience loses interest.

If there is a maximum time allotment for your presentation, be certain that you can complete your presentation at a relaxed pace and stay within the time frame. Increased confidence will come from knowing what you are going to say and from knowing that you will be able to complete it in the allotted presentation time.

6. *The 7 per cent rule.* After a 24-hour period, most people will retain about 7 per cent of what they have heard in a presentation. They will remember the noted three key points, and the rest will fade from memory unless the audience member actively seeks to shift more of the presentation from short-term to long-term memory by reading over notes, discussing the presentation, or spending conscious effort on remembering aspects of it.

    Why should this increase the confidence of the presenter? Because many speakers berate themselves for every stumble across a word, every point missed, and every perceived blunder in their presentations. The simple truth is that the audience probably was not aware of many of these supposed faults—they are not so familiar with the presenter's style as to be able to evaluate minute details in a presentation; and even more likely, the perceived mistake was so small it passed unnoticed by the audience. There is comfort in knowing that the audience's retention is much less than the speaker might expect.

7. *Know that the audience wants you to succeed.* Along with the 7 per cent rule, it is important to know that the audience wants the speaker to do well. People attend a presentation expecting to hear something of value. In general, audiences are enormously forgiving of the speaker as long as they get what is expected from the presentation. Know that the audience is 'rooting' for you—they are on your side. Just give the audience something; any presentation that is delivered with enthusiasm and sincerity will be well-received by an audience.

8. *Admit it.* If you are very nervous, tell the audience. They will react with empathy (most of them would be more nervous) and you can relax into your presentation. Being human and admitting your anxiety can create a strong connection between the speaker and the audience.

    It should be noted, however, that this technique will become less necessary over time. The more you present, the less likely you are to feel this nervous, and it will become unnecessary to share this with the audience. False humility (when the speaker pretends to share a frailty with the audience) is easily detected and will cause a loss of credibility.

Most important, a calm speaker is a *prepared* speaker. There is a profound difference between the anxiety one might feel over public speaking and the same feeling one has when woefully under-prepared for a presentation. Increased confidence comes from being prepared, planning for any problems or issues that might arise, and knowing the audience is on your side.

Chapter Review

This chapter began with a review of fear responses to public speaking and discussion of reasons why public speaking is so often feared. Various techniques for easing tension, relaxing, preparing, and building confidence have been considered here, so that the reader will be well prepared to let go of a fear response to presenting and to use the techniques provided to become a successful speaker.

Review Questions and Activities

1. What is your fear response? Do you tense shoulder or jaw muscles? Consider your past reaction to stressful situations and how you responded physically.

2. In the first section on the fear response, three methods are provided for calming the speaker. Practise each of these now.

3. List eight methods for increasing confidence as a speaker. Which one is most relevant for you?

4. What is the ideal energy state for a presenter? Why?

5. What is the 7 per cent rule? Why is this of interest to the speaker?

Chapter Ten

Speech Production

Presentations are more than saying words in front of an audience. The *way* we speak is also important. In fact, most of what an audience remembers is not what you said, but the way you said it—the tone of your voice and the non-verbal cues you provided with your presentation. This chapter considers the mechanics of speech: how speech is produced, and what shapes the sound of the speaker's voice. It also presents the elements of good speaking: the 'how you say it' parts of speaking. The chapter includes a number of exercises to improve the speaking voice. These should be performed while standing, since most of the presentations you make will require you to be standing in front of an audience. In addition, standing changes your performance state to a higher level of awareness. You can concentrate on voice production, gestures, and the use of space much more effectively when standing.

The way you stand is also important. You are likely familiar with the term 'core strength' in relation to exercise (for example, some forms of yoga and martial arts focus on the development of strength in the mid-point of the body for better results). Core strength also is needed for a good presentation. The power behind your voice comes from a point below your rib cage at the body's centre of gravity. In this central section of the body, more than 25 muscles work together through the abdomen, hips, and back to power performance. If you concentrate on working from this core area, your performance will have strength and intensity. Your audience will hear and see that you are ready to perform. When trying any of the exercises in this section, or when presenting, hold the following 'ready' body posture:

- Stand with your feet shoulder-width apart.
- Distribute your weight evenly on each foot, focusing on keeping your weight forward to the centre and front of your feet.
- Centre your legs and hips above your feet so you are working from a steady base.
- Focus on your core area and on keeping core strength: your voice and the power of your presentation emanate from the core area.

- Pay attention to your breathing. As you exhale, release your shoulders, relax your neck, and unclench your teeth.
- Hold your shoulders back, not unnaturally so, but straight and square with your core.
- Hold your head upright, with your chin parallel to the ground.
- Feel as though you have an imaginary string attached to your solar plexus. The string pulls upward in front of your throat, through your head behind your eyes and nose, and exits at the crown of your skull. The string is tied to some infinite point far away in the universe. You can feel your body pulled gently in line along this string, causing your shoulders to remain in a relaxed and squared position.
- Your arms are hanging loosely at your sides, with your thumbs resting on the outside seam of your jeans. Your hands are still.

Practise this 'ready' posture frequently to give yourself the ability to quickly assume it whenever you have the opportunity to present.

Vocal Production

As infants, we produce sounds involuntarily, and quickly learn that we can generate a positive reaction if we make certain sounds at certain times. We learn that there are patterns to sounds, and start to detect repetition in spoken words. Over time, we learn that these patterns and sounds have meaning: we learn the differences between the terms that describe the female and male parental figures and the canine member of the household.

Researchers have yet to come to agreement on the number of words in the English language, but most would agree that we quickly acquire a working vocabulary by age four or five, then at some point realize that words are symbolic of meaning and begin to add depth of meaning to our vocabulary.[1] We continue to acquire words throughout our lifetime by absorption (hearing or reading new words and attaching meaning to them) or by actively working to expand our vocabulary.

Over time, we also develop speech patterns—what the voice sounds like to listeners. In most cases, the sound of one's voice develops with little conscious thought. It is shaped in part by the physical structure of the parts of the body involved in speech: the core muscles, lungs, throat, mouth, and the resonance chambers in the head. As sound is produced, it reflects the condition and shape of these physical components.

However, the sound of your voice is shaped by other factors. The first is imitation. Often, members of a family will sound the same on the telephone to a listener, sometimes to the point where the speakers cannot be differentiated. What the listener is hearing is the absorption of similar speech patterns by members within a closely related group. As infants, speech patterns are noted and replicated, and this process of replication develops norms for sound production and ways of speaking. More widely, imitation in speech among members of a social group leads to the production and retention of accents. An accent is an entirely acquired element of speech that has no relationship to the physical structures of the body; it is absorbed through listening and also taught by parents or teachers who correct a child's pronunciation.

The second factor shaping the sound of your voice is deliberate manipulation of speech patterns. Actors assume accents and speakers change their vocal patterns depending on the listeners. These changes are conscious: the speaker manipulates the mechanics of speech to produce a different sound, either over a short time period (the length of the play or part in a movie) or over a lifetime (amending an accent to assimilate or to dissociate within a population).

First, consider the two mechanical components of speech that have the greatest influence on the sound of the speaker's voice: breathing and resonance.

Breathing

The autonomic nervous system keeps us breathing, and this breath is used to activate speech. Air is drawn into the lungs through the nose and mouth, generally in an amount that relates to the person's level of activity at that moment—reading a textbook requires less air than intensive exercise. At this moment, it is likely that you are 'shallow breathing' and using only a small portion of your total lung capacity.

On exhalation, air travels from the lungs across vocal folds in the trachea; we activate the muscles around these vocal folds to cause them to move and the vibration created is what gives us 'voice'. Strength in the voice is gained by moving the airflow more quickly and more powerfully through the body and also by using the body as a sounding board (see the discussion on 'resonance' below). The tongue, teeth, hard palate, soft palate, and lips then shape sound into words. With sufficient volume and good articulation (the way the words are formed and pronounced), the listener is able to understand the speaker's meaning and intent.

In presenting, we use the voice differently from the way we use it in conversation. The most important difference between speaking and presenting is in the way breath is used to power the voice. Economy of breath is needed to create a steady and powerful flow of air across the vocal folds; try this exercise.

- Place your hands on your waist, just under your rib cage.
- Breathe in and out through the nose with full breaths, soundlessly.
- Concentrate on filling the lungs entirely before breathing out.
- Feel your hands move in and out as your lungs fill and empty.
- Concentrate on using your core strength to assist your lungs in moving air.

Now, take a full breath and count out as long as you can, 1, 2, 3, . . . until you have no breath remaining. You may find that you need to force out the last few numbers as you run out of breath. Take a deep breath again, and without forcing the breath or collapsing into a shortness of breath, count out again, 1, 2, 3, . . .

Did you manage to count to a higher number? Try again, using economy of breath to extend the counting range and using your core strength to maintain power in your voice. Your final number should be as clear and powerful as your first number.

This is a good exercise for understanding the connections among your lungs, voice, vocal apparatus, and brain in producing speech. It is also a good exercise for learning economy of breath for presenting. You can speak only as well as you breathe. Control your breath, and you can control your speaking voice. Also, better breathing reduces tension in the neck and shoulders that can inhibit your best natural voice. Regular practice of the exercises in Box 10.1 will help to improve breathing and release the tension that can accumulate in the body. The added benefit is increased oxygen supply, which will lead to greater alertness and increased calmness.

Box 10.1 Breathing Exercises

Ha ha ha

One method for improving the flow of breath is to say 'ha'. This also helps to improve economy of breath as the word 'ha' is one where speakers tend to expend a great deal of breath and energy. Work to economize the air needed to say each 'ha', speaking in a clear and powerful voice. Also, do not overly stiffen the chest and shoulders while doing breathing exercises. Keep the upper body tall, with shoulders square to allow the lungs to work at full capacity.

How much breath?

Think about how much breath you will need to count from one to five. Draw in a breath, count out to five in a normal conversational voice, and determine if your estimate was accurate. Repeat until you are certain on how much breath is required to complete the phrase. Now do the same, but count to 10, 20, 30, . . . only drawing in the required breath.

S and F

Take a full breath, and then breathe out in a quiet, sustained manner, economizing the outgoing breath while alternating between making an 'S' sound and an 'F' sound. Make the sound perfectly even in tone and volume, and work at it until the sound can be maintained evenly for the entire exhalation.

Resonance

Resonance is the reinforcement of sound and the way the body projects sound. The sound produced by the vocal folds is small; however, as the sound moves through the hollow spaces in the vocal tract and head, the sound is amplified. That is not to say that you have an empty head; more specifically, your nasal passages and mouth are the resonator cavities in your head. Your throat is also a resonating chamber.

Sound resonates better against hard surfaces and through hollow spaces. Primarily, the sound of your voice is based on how your voice resonates through the oral and nasal cavities on the front of your face. Sound also bounces off the back of your teeth, the ridge behind your teeth, and the hard palate on the roof of your mouth. You need to push the sound forward to the front of your face and these hard surfaces for the sound to be powerful.

A resonant voice is one that uses this natural amplification of sound to full advantage. The resonating cavities are kept fully open, and sound travels freely from the vocal folds. The voice is clear and full. A listener is attracted to a resonant voice, because the voice sounds powerful and complete.

Try the following yawning and humming exercises to improve resonance:

- *Yawning.* The easiest and most productive way to open the resonating chambers is to yawn with your mouth wide open. Open the throat as you yawn, as well. This carries the side benefit of causing you to expand your lungs to full capacity and injecting needed oxygen into your body. Note the difference in throat tension before and after you yawn. A yawn is also an effective way to relax tense throat muscles.
- *Humming.* Start to hum, and then push the sound to the front of your face. Try to make the mask of your face (the bones and support structures) vibrate from the power of the hum. Concentrate on keeping the sound forward. Pushing the voice forward fully engages the resonance chambers, while holding the voice back in the throat makes the voice sound muffled. The difference is obvious if you consider the sound of your voice when you have a head cold (therefore, the resonance chambers in the head are fuller) compared to when you are healthy. The voice sounds compressed and stifled, both to the speaker and the listener. Resonance is compromised and the voice loses warmth and depth.

Practical tip. These exercises can be performed any time. While it is better to concentrate on the exercises and perform them while standing, these exercises can be practised in any position. Try yawning and humming as you commute to work or university.

Breathing and Resonance

Speakers sound different because the physical aspects of their bodies involved in speech production—the core muscles, lungs, throat, and the resonance chambers in the head—are different. Males tend to have deeper voices than females because testosterone lengthens the vocal folds, and individuals who practise whole-lung breathing (as do professional singers) tend to have more powerful and more resonant speaking voices. Voices change with age, as well. As one ages, the voice becomes less resonant, and lessened breath control can add a shaky quality to the voice.

> ## Box 10.2 Improving Resonance
>
> To improve resonance, say the following, first concentrating on pushing the humming sound to the front of the face, then letting the resonance push out through the final syllable:
>
>   MMMMMMMMMM mah
>   MMMMMMMMMM mee
>   MMMMMMMMMM my
>   MMMMMMMMMM moo
>   MMMMMMMMMM mew
>   MMMMMMMMMM maw
>   MMMMMMMMMM may
>   MMMMMMMMMM mo
>
> Work through the exercise three times, noting improvements in resonance.

Without surgical intervention, it can be difficult to change the breathing or resonance apparatus to affect the sound of the voice. However, with the implementation of good breathing techniques and improvements in resonance, the speaker can create a more powerful voice that a listener hears as more attractive and authoritative, a voice that sounds more powerful and vibrant. Breathing and resonance are two mechanical aspects of speech, but the sound of the speaker's voice is also shaped by the elements of speech that provide individuality and meaning—these elements affect 'the way you say it' and the conscious shaping of these elements of speech is critical to a successful presentation.

The Elements of Speaking

Good breathing techniques are the first step towards shaping the sound of the voice, followed by developing depth in the voice by exercises to increase resonance. On these foundations to speech rise nine elements that shape what the listener hears. Each is considered below, along with techniques for improving the speaking voice.

Pitch

Pitch is the frequency of vibrations of your voice. A high-frequency vibration has a high pitch; a low-frequency vibration has a low pitch.

The pitch of your voice is partially regulated by the mechanics of speech—the physiology of your vocal folds and the speed at which air is forced over the vocal folds, creating tension. Generally, males, with thicker vocal folds and elongated voice boxes, have lower-pitched voices, while female voices generally are higher-pitched.

Box 10.3 Elements of Speech

The elements that form the way we speak include:

1. *Pitch*: frequency or vibration of the voice
2. *Pace*: rate of speech (a fast pace or a slow pace)
3. *Tone or timbre*: the warmth or colour in the speech that provides meaning
4. *Emphasis*: speech patterns that stress words or syllables
5. *Inflection*: related to pitch, patterns of changing the pitch of the voice
6. *Pause*: related to pace, the use of silence in speech
7. *Volume*: managing the loudness of speech
8. *Projection*: moving the voice into the audience
9. *Articulation*: the way words are shaped and syllables pronounced

Humans have the ability to vary greatly the pitch of their voices—sometimes as much as two octaves in a speaking range. Generally, the pitch of a presenter's voice will rise with excitement. The more enthusiastic you are, the more likely that you will be speaking at a higher pitch. Consider the voice of World Cup soccer announcers; a rise in pitch is typical when a goal is scored. The pitch of the voice also typically lowers when one is speaking of something very serious. Along with a quieter voice, the speaker will use the lowest notes of the voice to deliver an important message.

As listeners, we innately correlate a higher-pitched voice with excitement and a lower-pitched voice with seriousness, relative both to the vocal range of the speaker (if we are familiar with that voice) and in comparing speakers to one another. As we become familiar with a voice, we learn the average pitch of the voice and then derive meaning from variances in its pitch. Therefore, varying the

Box 10.4 Pitch

To find the pitch range of your voice, start by counting from one to twenty in your normal pitch. Next, start at 'one' but consciously lower the pitch of your voice with each number. At some point, you will reach a pitch that is no longer easy or comfortable to produce.

Do the same exercise, counting from one with an increasingly higher pitch. Again, work to find the highest pitch that remains comfortably within your vocal range. Guard against shifting into a singing voice from a speaking voice.

Your vocal range can be expanded with practice, but the issue for most people is that they are not using the pitches already within their range. Listen to your voice and ensure that the pitch matches your intended meaning.

pitch of your voice can be highly effective in delivering a message to an audience. For points that are critically important, speak with the lowest notes of your voice. To raise interest use the highest pitch that is comfortable and within your speaking range. Work to consciously raise and lower the pitch of your voice to match your speaking voice to the meaning intended in your words.

Depending on your speech topic, you may want to consider the average 'note' of your voice. Is this a serious speech or a sales pitch? Are you trying to convince the audience of a critically important point, or persuade them to act now, while supplies last? With practice, you will be able to manipulate your natural speaking pitch. Practise until changing the pitch of your voice becomes easy and natural while you are speaking.

Also, vary the pitch of your voice. The overuse of high-pitched with fast-paced speech will make your presentation come across like a late-night TV infomercial, which is unlikely to be the intended effect in most academic and professional environments. However, a monotonous, low-pitched voice, while serious, may have the unintended consequence of putting the audience to sleep. Use the low notes and high notes in your vocal range, varying the pitch to match the meaning of your words. As with the application of most techniques, also be aware of potential overuse: too much variance in pitch sounds 'sing-song' and contrived, and has the additional consequence of not relating pitch to meaning.

Work to develop your vocal range and use pitch to your advantage. Having power over pitch and varying your pitch will add credibility and excitement to your presentation.

Pace

The average speaking pace for a North American English speaker is approximately 125 to 160 words a minute. How does pace impact the audience's understanding of the speaker's words? At first glance, one might think that speaking at a slow pace is the best way to impart meaning. However:

- At too slow a pace (less than 100 words per minute) the listener loses the ability to keep sentence structures intact. Words lose meaning, and the listener will be easily distracted from the presentation.
- At too fast a pace (more than 190 words per minute) the listener is unable to assimilate meaning from the words. At first, the listener will intently attempt to follow the speaker, but interest will quickly be lost and the listener will no longer be effectively engaged in the presentation.
- The most effective rate is an average of 140 words per minute. The listener is able to hear and assimilate the words, and may be able to take notes on the presentation, increasing the likelihood that key points will be remembered.

Generally, speakers who speak at an average or faster pace are viewed as more enthusiastic and more credible than speakers who deliver a presentation at a

Box 10.5 Pace

Read the following Winston Churchill quotation, first at a normal conversational pace, then very quickly, then very slowly, with particular emphasis on the last five words.

'Let us therefore brace ourselves to our duties and so bear ourselves that if the British Empire and its Commonwealth last for a thousand years, men will still say, "this was their finest hour".'

Hear how differently the sentence sounds when the pace is varied.

slower than average rate. While machines can measure rates of speech, an easier method to evaluate the pace of your own speech is to find a selection to read (of about 200 words) and see how far you read aloud in one minute (using your regular rate of speech and a normal conversational tone). Test yourself again, using the rate of speech you would use in a presentation. Repeat this exercise until you get a sense of your own rate of speech, both in a conversational sense and as a presenter.

Do you normally speak in a fast to very fast range? You may want to slow your speech to ensure that listeners have sufficient time to process your words. This is critically important for speakers who tend to increase their pace of speech when presenting. Do you normally speak at a slow rate? Consider speeding up your speech to appear more excited and enthusiastic about your topic.

Within a presentation, vary the pace to impart meaning. Speak at a slower rate if you want the audience to hear that you are 'thinking' about your statement. For less critical information (like a list), speed up your speech. Note that a higher rate of speech may sound more animated, but the audience still must be able to hear what you are saying, process it, and understand it.

The speed at which you speak can be varied to provide emphasis. Consider the most important point of your presentation—it may be more powerful to deliver . . . it . . . slowly. Learn to hear your own rate of speech, and be able to adjust this rate as you are presenting to create a varied and enjoyable performance.

Practical tip. There are many websites and sources for speeches—YouTube and Google videos are just two of the myriad places on the web to view speeches by both good and not-so-good public speakers. Seek out the videos of recognized public speakers (politicians, business people, personalities) and critically evaluate the speed of their performances as well as all other aspects of performance. What works? What doesn't? What do you take away from these performances?

Tone or Timbre

More than any of the other elements of speech, the tone or timbre of your voice provides the emotion and colour to your presentation. Tone or timbre is what people are speaking of when they ask, 'What did you mean by that?', when your words are innocuous, but your tone is not.

The inability to hear the 'tone' of the writer's voice is one of the reasons why e-mail can easily be misconstrued by the reader. Words that sound harmless in the mind of the writer can be read in a completely different tone by the reader. Consider the statement, 'Nice job at the meeting.' What two meanings could you construe by changing your tone?

Is the tone of your voice warm and friendly when presenting, or devoid of emotion? If your tone is weepy or fearful, your presentation may lack credibility in a business or professional situation. Conversely, if your tone is flat and cold, you may not be viewed as believable when you are sharing personal reminiscences or information.

Adjust the tone of your presentation to suit the audience and your material. Ideally, to sound most natural, the tone of your voice must reflect the way you feel about the topic. Is it interesting? Terrifying? Does the emotional quality of your voice match your words? For example, say, 'I am so happy to see you!' while speaking with an angry tone in your voice. The incongruence between your words and tone will increase the likelihood of being misunderstood by the audience.

> ### Box 10.6 Tone or Timbre
>
> Many speakers are reluctant to record their speech because they do not enjoy listening to their own voice. Your voice sounds different to you in a recording because you are not hearing it through the resonators within your own body as well as through your hearing apparatus. Your voice sounds different because the experience of the resonators are removed from the recording. However, this is a useful means of evaluating whether the tone of your voice fits your intentions. As you listen to your voice, think of the words that describe it. Do you sound excited, engaged, and interested in the topic, or monotonic and bored? Do you sound angry? Is the tone apologetic? Consider whether the tone matches your intent, and if it does not, consciously work to change your delivery.

Emphasis

The way you 'hit' your words with tone and power creates emphasis in speech. Normally, the most important words are emphasized, creating focus in your presentation (like a verbal underlining of the word or phrase). Emphasizing a word gives it greater significance than the words surrounding it. Consider the difference

Box 10.7 Emphasis

An acting technique for finding the right meaning of lines uses emphasis. The actor will emphasize different words in a sentence to see if it creates a new nuance of meaning. Perhaps the best-known example is Robert De Niro's delivery of a pattern of changing emphasis in the movie *Taxi Driver*.

You talking to me?
You *talking* to me?
You talking *to* me?
You talking to *me*?

The meaning of each question changes, depending on the placement of emphasis.

between these two sentences when the emphasis is shifted from the first, to the second, to the last word:

He saw me.
He *saw* me.
He saw *me*.

In the first sentence, the listener hears that the speaker has placed importance on the person who saw her (we'll have to hear more to find out if this is positive or negative). In the second, the concern is that the speaker has been seen (perhaps she was planning on missing class, but was seen by the professor at the coffee shop). In the third, the emphasis is on the speaker: her greatest hero waved at her in the airport. The point is that changing the emphasis changes the meaning of the sentence.

If you pronounce every word in a sentence with equal emphasis, the listener loses the ability to distinguish the difference between what is important and what is less so. This form of speaking can be highly emphatic in a short statement if used infrequently, but it can become very tiring and lose meaning if overused. The voice can become robotic, and the listener cannot derive meaning from the patterns of emphasis.

In addition, if you place emphasis on the wrong words in a sentence, the meaning of the sentence can also be lost or altered. To hear how changes in emphasis change the meaning of a sentence, read the following aloud, putting emphasis on the italicized word:

I was born in Australia.
I *was* born in Australia.
I was *born* in Australia.
I was born *in* Australia.
I was born in *Australia*.

The change in emphasis entirely changes what is heard by the audience. When presenting, ensure that the emphasis used matches the message you are delivering in your presentation.

Speech elements such as emphasis can be misunderstood, thereby changing the meaning of the sentence if your words are being translated or transcribed. If you are speaking in a venue and know in advance that your speech will be reproduced either as an audio translation or on paper, read through it before the presentation to be sure you have not set up statements that likely will be misinterpreted.

Inflection

Inflection is related to the pitch of your voice, but instead of focusing on whether you speak in a low pitch or high pitch, inflection considers how this pitch is altered over words or phrases. Again, the meaning of words changes with the way you say them. Consider the word 'oh'. If said as a question, along with a raised eyebrow, the speaker's inflection would indicate disbelief. If said as a flat statement with no change in pitch, it could be viewed as neutral, or perhaps carry connotations of disappointment. If said in a circumflex manner—that is, starting on a low note, shifting to a higher pitch, then returning to the low note—the meaning changes again to connote 'I told you so.'

There are four types of changes in inflection:

1. *No change*. The voice maintains the same 'note' over the entire sentence (sounding robotic).
2. *Inflection up*. The voice rises at the end of the sentence, sounding like the speaker is asking a question. This inflection makes sense if you are actually asking a question, but is heard as timid and uncertain by the listener if it is overused and will quickly diminish your credibility as a speaker. This speech habit, called 'up speak', is unfortunately becoming more prevalent (especially among teenaged females). The speaker falls into a pattern of ending every sentence on a question (We went to the mall? And I saw this sweater? And I might get it? But I need some money?). The addition of the word 'like' to every sentence will further detract from the meaning and credibility of speech. While this may not be an issue at the mall among the teen's peer group, it is an issue from an academic or career perspective. The speaker does not sound credible. On a positive note, it is possible to alter a speech habit like 'up speak' in a fairly short time, if the speaker is motivated and interested in adopting new speech habits.
3. *Inflection down*. The voice falls at the end of the sentence. A falling inflection is correct for most sentences, although it can be misused if a question is ended with a down inflection. The speaker can sound angry or overly emphatic, and the listener will be unsure of the meaning of the sentence. Consider the statement, 'Is this the final version of your paper?' Ask it first as a question, and then with a downward inflection. From the listener's perspective, the meaning of the two sentences is entirely different.

Box 10.8 Inflection

Say the following statements, with the pitch of your voice following the four inflection changes shown by the arrows:

Statement	Type of Inflection Change	Vocal Direction
Hello.	No change	→
	Pitch up	↗
	Pitch down	↘
	Pitch varied	↺↻
Yes, I will.	No change	→
	Pitch up	↗
	Pitch down	↘
	Pitch varied	↺↻
No chance of that.	No change	→
	Pitch up	↗
	Pitch down	↘
	Pitch varied	↺↻
Is that what you mean?	No change	→
	Pitch up	↗
	Pitch down	↘
	Pitch varied	↺↻
Stop doing that.	No change	→
	Pitch up	↗
	Pitch down	↘
	Pitch varied	↺↻

Note the change in meaning with the changes in inflection. Again, be certain that the way you say your words matches your intent.

4. *Inflection up and down (or down and up, or any variation where the pitch changes in more than one direction).* This type of change can be used to show surprise, sarcasm, doubt, or uncertainty.

Pause

Never underestimate the power of pauses in presenting. Consider the statement, 'I will now let you in on the *greatest* secret in public speaking . . .' followed by a lengthy pause. The pause allows the audience to refocus and pay closer attention on the statement that follows the pause.

The use of a pause can also make it appear that the speaker is thinking and adjusting the presentation while speaking. The audience sees that he or she is not simply reading, but tailoring the presentation to the audience present, again adding to the presenter's credibility.

Box 10.9 Pause

A pause can be used to:

* Attract attention.
* Refocus the audience.
* Control the behaviour of distracting individuals.
* Bring a presentation to a key and final point.

Use a pause as a focusing technique—anything that comes after it will appear more important.

A pause is also related to timing: a pause before saying the punch line of a joke heightens the anticipatory tension in the audience to produce a more intense reaction. A pause can also be built in for the purpose of dramatic effect—'and the murderer is . . . the professor!'

In your next speaking opportunity, try using some pauses. Complete a sentence, and then stop for 10 seconds. This will likely feel like 10 minutes, but do not allow yourself to continue. When you do begin to speak, you will find that the audience has refocused, you have allowed yourself an opportunity to collect your thoughts, and the remainder of your presentation will be more powerful.

A pause also is a useful technique when attempting to stop side conversations between audience members. If you find that you have audience members chatting while you are presenting—and this is distracting to you and the listening audience—pause. Without being overly aggressive, look in the direction of the talkers. Within seconds, the audience will focus on them and the chatterers will likely self-police into quietness. In a world surrounded by constant noise and interference, the use of the occasional pause will command extreme attention. Use it to your advantage. It is the greatest secret in effective public speaking.

Volume

The volume of your voice is critically important. No single factor loses the interest of the audience more quickly than a presentation that cannot be heard. In fact, after asking the speaker to turn up the volume or speak more loudly, the frustrated audience will often turn on the speaker, creating a negative experience that could have been avoided by increasing the volume of the speaking voice.

Conversely, a presentation that is too loud, either because of the speaking volume of the presenter or too powerful a sound system, also is disturbing to the audience. While a high volume is acceptable in some circumstances, most spoken presentations do not benefit from being too loud. The audience will react negatively, and may even leave the venue.

The volume of a presentation can be controlled. First, the speaker has control over the volume of her or his voice. With projection (see below), the speaker can push his voice more powerfully into a space. Full lung breathing and the use of core strength will add to the volume of the voice without increasing shrillness. In addition, projecting the voice protects the throat muscles and speaking apparatus, which can be damaged if the speaker shifts into shouting to be heard.

In addition, many presentation spaces are equipped with microphones. As a rule of thumb, if there are likely to be more than 70 people in attendance or if the venue is built to hold this capacity, a microphone may be needed, depending on the acoustics of the space and the ability of the speaker to project her or his voice. While having a microphone may at first seem to be of benefit to the speaker, there are many issues with using amplification that must be considered:

- *The type of microphone.* Is it a lapel pin or a headset? Handheld or attached to the lectern? Some speakers prefer to move about the presentation space and will need a wireless system, while others like to hold notes or cards and cannot manage a microphone as well. Will you be comfortable using the microphone system that is available?
- *Range.* Some are carefully tuned to capture only the speaker's voice while others will amplify every ambient sound, like paper-rustling or throat-clearing in the audience. You may need to increase the distance from other potential sources of noise
- *Sensitivity.* If the microphone is highly sensitive, every 'p' or 'b' spoken will reverberate with the audience like a gunshot, and the speaker's breathing can sound like Darth Vader.

If you are invited to give a presentation, be certain to inquire if a sound system is available and the type of system that will be used. Ask how to use it. If the microphone must be held at a certain distance from your mouth to be effective, find out what this distance is. With a handheld microphone, too many presenters start out correctly, then lose this position as they move into the presentation. The audience can no longer hear the presentation and will rapidly lose interest. With a fixed microphone, many presenters will neglect to adjust it to their height and will either hunch over the microphone or be straining to reach the required height—either way, credibility is lost as the presenter appears uncomfortable.

Even if a technician is on hand for managing a sound system, the audience sees the speaker as responsible for the quality of the presentation. If possible, given the size of the venue and the crowd, try to rely on your own ability to project your voice rather than on amplification, as there are far fewer things that can go wrong.

As the speaker, you have a great deal of control over the volume of your presentation. When you check out the venue before your presentation, note the size of the room and any factors that might impact the ability of the audience to hear your presentation. It is important to ensure that doors and windows are closed and noise in adjacent rooms is kept to a minimum. Find out if fans will be running as part of the heating/cooling system, and test their volume. Look for other mechanical or human-caused noise (e.g., the group meeting in the next room) and determine if there are ways to reduce it.

Box 10.10 Volume

Say the following sentence, 'The meeting is about to begin':

- To someone standing right in front of you.
- Conversationally to a small group.
- To a classroom at full capacity.
- To a gymnasium full of people.

Work to push your voice using strength from your core, not from your throat. Create greater energy and your voice will be heard at a greater distance.

Learn to tailor the volume of your voice to the size of the room—the audience does not want to be shouted at, but they do want to hear your presentation clearly. As you increase the volume of your voice, take care that you are not straining your throat to shout at the audience; instead, use the power of projection (outlined below) to ensure that you are heard.

Varying the volume of speaking adds interest to the presentation and keeps the audience focused. Generally, the volume of a speaker's voice increases with excitement—this can also be used to the presenter's advantage. And don't neglect the selective use of a low volume; a stage whisper can be used to pull the audience in, although the speaker must be certain to maintain sufficient volume for all audience members.

Projection

As stated before, there is little point in presenting if the audience cannot hear what you are saying. Whether or not a microphone is available, it is up to you to ensure that you can be heard throughout the venue. Even without a microphone, you still have great ability to project your voice to ensure that you can be heard.

Projection is different from volume: it is not shouting at the audience or straining to speak more loudly. Instead, projection is using core strength to 'push' the voice to the edges of the room. Along with this added energy, the presenter must speak clearly and slightly raise the pitch of the voice if speaking in a large venue.

To project your voice, 'think' your voice to the corners of the room—keep your chin parallel to the floor and use core strength to create better breath control. The throat should not feel strained because all the power behind your voice is coming from your centre. Speak only as far as you need to: if you are speaking to a small group clustered close to you, project to them. If you are speaking to a classroom of students, project out to the corners, noting, as well, that greater projection is needed in a crowded venue as sound is absorbed by the audience and the audience members themselves will be coughing and shuffling papers.

Next, consider your posture. Are you standing in a 'ready' position? Be certain that your chin is raised and that you are speaking directly towards the audience and not to the lectern or to the screen behind you. Inhale, filling your lungs with more air volume than you would normally need for standing. This extra oxygen will power your voice, keep you calm, allow your brain to fire rapidly, and give energy to your presentation. The breath will be drawn in from your core, not from the top of your lungs. Use this core strength as the foundation for powering your voice out to the corners of the room.

Box 10.11 Projection Exercises

Calling Away
Picture someone 30 metres away, and call:

> Hello! Hello!
> You forgot your keys!
> Here Rex! (or your favourite dog name)

Feel the tension in your core as you make your voice louder. Now bring that muscle memory to a presentation space, and read a selection of your choice, pushing your voice to the corners of the venue.

Good projection feels powerful. Poor projection (based in the upper lungs and throat) will quickly strain the voice and may even become painful. Practise projecting into a variety of spaces until varying your projection becomes normal.

Articulation

Articulation is the way you say your words—the way you pronounce your consonants and vowels, complete the syllables in words, and tie words together. Well-articulated speech is clear, easy to understand, and correct; the listener can understand both the word and the meaning without difficulty.

In poorly articulated speech, the speaker may slur words, trip over syllables, miss out on the hard consonants that shape speech (like the 't', 's', and 'd' at the end of words), or mispronounce words. It is exhausting to listen to poorly articulated speech, and an audience will quickly lose interest if they cannot understand what the speaker is saying.

How can a speaker improve articulation? First, be sure to actually move your mouth when you speak, especially when presenting. Often, nervousness is felt in the jaw and throat, and muscles tense when you are feeling stressed about a presentation. You start to speak with minimal jaw movement, like a ventriloquist. Try to read this paragraph aloud without moving your jaw. Unless you are a former vaudeville performer, it is likely that your speech will become harder to understand. Now read it again with full movement of your jaw, mouth, and lips. Your speech will be distinctly clearer.

To loosen the throat and mouth muscles (referred to as the 'embouchure'— all the muscles, ligaments, and parts of the body from the upper lungs and up, involved in the production of speech), simply yawn. Yawn widely, stretching at

Box 10.12 The Greatest Warmup Ever

The single greatest warmup exercise to improve articulation is:

Ma Me O My O Mu

These syllables cover all vowel sounds and offer a rapid way to warm up for a presentation.

Say the syllables over and over, exaggerating your facial movements. Vary pitch and tone while you are speaking to improve your vocal range at the same time. Combine with exercises that improve other elements of speech, such as pitch or tone, to improve two elements of speech at once.

the same time to loosen other muscles. Feel how a large yawn opens up the mouth and throat, relaxing the muscles and allowing the speaker full movement of the jaw. Ideally, this exercise should be done before the presentation and not in view of the audience.

A second way to improve articulation before a presentation is to run through a number of tongue twisters or syllables that 'warm up' your embouchure. Say each of the following five times, stretching your mouth as you speak to gain maximum benefit:

> *Bee mee bee mee bee mee bee mee*
> *Baa paa baa paa baa paa baa paa*
> *Knapsack straps*
> *Unique New York*
> *Greek grapes*
> *Cheap ship trip*

If you are presenting first thing in the morning or on a day when you have not spoken very much, be sure to warm up your voice and embouchure before presenting. Athletes insist on warming up before a sporting event, but presenters sometimes fail to give themselves the same consideration when presenting. You need to warm up. Sing along with the radio as you commute, say tongue twisters or nonsense syllables, or practise your presentation: your objective is to arrive at your presentation fully warmed up and ready to speak.

How to Change Your Voice

While it is true that the shape and positioning of your vocal apparatus are largely innate, the way you speak is an entirely learned behaviour. You were instructed, and largely learned by example, how to pronounce words and syllables and how to form sentences. You learned, through observation and imitation, appropriate ranges of pitch, pace, and volume in your voice. You mimicked the sounds that brought a positive response, and these habits became ingrained. Once you established these habits, your speech became your voice.

You may have an accent that identifies your place of origin. This is a learned characteristic. There is nothing biological or instinctive about this accent, nor is it inherent to you as a speaker. If you had learned to speak in another place on the planet, your accent—not to mention your mother tongue—would be different.

It follows that if the way we speak is learned, it can be changed. While physiological characteristics or damage to the voice-production apparatus, such as a deviated septum or extensive orthodontics, can cause a speech problem, most of the issues with our speech are related to habits rather than to actual physical issues.

The first step in changing your voice is to become aware of how your voice sounds. While most people find it difficult and even unpleasant to listen to their own voice, it is important to be able to step back and critically examine vocal production to see if there are aspects requiring improvement. Generally, people are surprised at the sound of their own voice and will be interested in learning techniques to make the voice more powerful, resonant, and interesting.

One issue that must be addressed at the outset: working to improve the speaking voice is not phony or insincere. A significant percentage of the population will pay a monthly fee for a gym membership and work to manipulate the structure and form of their body shape, but find the idea of manipulating the sound of their voice offensive and somehow less than genuine. However, just as you may choose to alter your body shape, you can choose to shape the sound of your voice through focused exercise.

In analyzing the speaking voice, several characteristics may be examined:

- *Volume*: the natural volume may be too loud or too quiet.
- *Rate of speech or pace*: the pace may be too fast or too slow (technology is available to measure words per minute).
- *Pitch*: the natural pitch of the voice may be too high, too low, lacking in variation, or lacking in range.
- *Tone*: the tone may be harsh, flat, shrill or warm, varied, or modulated. Overall, does the tone of the voice 'fit' with the image of the presenter and the content of the presentation?
- *Articulation*: are words pronounced with clarity in speech and inflection?

If you find that your performance could be improved in any of these characteristics, a good place to start is with the exercises in this chapter. Select one or two exercises specific to your concern and practise daily. Be assured that it is not false or pretentious to attempt to change your voice, no more than would be an attempt to change your physical appearance by getting an attractive haircut. Your voice is one of the ways you are evaluated as a professional: clear, crisp articulation, well-modulated tones, and a rate of speech that fits your profession are important. Sometimes, it can be difficult to hear one's voice objectively: working with a professional vocal coach is a recommended approach for changing your speech habits.

Chapter Review

This chapter examined the mechanics of speech: how air is transformed into sound, how the elements of speech shape the listener's understanding of the speaker. By understanding how speech is produced and the various elements of speaking, the speaker is better able to improve the speaking voice and become a highly effective presenter.

Review Questions and Activities

1. What are the elements of speech? List and explain.

2. What is the single greatest warmup exercise? Why is it so effective?

3. Record your voice, reading aloud a passage from a book. Give an objective review of your voice. What characteristics stand out? Are you surprised at the sound of your voice? What would you recommend to someone else interested in changing his or her voice?

4. Stand up and stand 'ready' as described in this chapter. Break the stance by sitting down, then stand up and move into the correct posture. Do this 10 times, until the ready stance is easily obtained upon standing.

5. Practise the breathing and resonance exercises outlined in this chapter.

Chapter Eleven

Non-verbal Communication

Try this exercise when you next have the opportunity: tell someone 'It looks like a great day today!' while shaking your head 'no'. Ask the person, 'How did you interpret my comment?' Invariably, the individual will say that you did not seem to think the day had the potential to be great, that you seemed sarcastic, possibly insincere, and rather disagreeable. This reaction would be based on your *non-verbal communication*: your 'body language', or the facial expressions, gestures, mannerisms, and stances that you use to transmit feelings. These non-verbal cues have more of an influence on the audience's perception of your speech than your actual words: in fact, research shows that an audience's interpretation and impression of your speech is based:

- 7 per cent on the words you say;
- 23 per cent on your tone (how you say the words);
- 70 per cent on your non-verbal communication.[1]

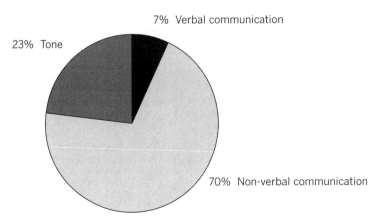

Figure 11.1 The Importance of Non-verbal Communication

The Forms of Non-verbal Communication

As noted in the simple exercise above, when non-verbal cues are disconnected from words, the speaker appears to not be speaking truthfully. One does not have to be a skilled interrogation agent to observe these cues; as humans, we have an amazing ability to detect the slightest incongruence between words and actions. A false smile, lack of eye contact, and fidgeting feet or hands are non-verbal signals that make a presentation less believable.[2]

In addition, the evolving field of 'micro expression analysis' examines the uncontrollable, fleeting expressions that wash over the faces of people trying to control or hide their thoughts and emotions. While some individuals have greater control over their facial expressions than others, such as those able to hold a 'poker face' or talented actors, current research in the study of non-verbal communication states that even those with the ability to control their facial expressions can do so only at a macro level; the face still provides 'micro expressions' that are measured in tenths and hundredths of a second that reflect the true feelings of the individual. Audience members paying close attention to the presentation will be able to detect these micro expressions, and the presentation will not be viewed as credible. Thus, even if your words are true, remember that the audience will place greater weight on their interpretation of your non-verbal actions. Consequently, it is imperative that your non-verbal communications fully align with the message you are delivering to the audience.

To begin, first, maintain a ready stance, as discussed in the previous chapter. Slouching causes the compression of the lungs, tension in the throat, and makes your voice sound constricted. In addition, the speaker appears nervous and unconfident. Find that place of good posture, being relaxed and alert, ready but not tense. Keep your feet planted on the floor, moving them only if you are actively moving about the stage. Do not shift your weight or lift one foot or the other; you will appear fidgety and unbalanced.

Box 11.1 Relaxing Hands and Arms

To assist in creating relaxation in your arms, pick up a medium-sized text-book in each hand. Hold your arms straight out at your sides, parallel to the floor, until your arms get shaky or tired. Set down the books and then let your arms fall to your sides. Note the feeling of relaxation. Use 'muscle memory' to allow your arms to fall loosely at your sides in a relaxed state.

To relax your hands, alternate between clenching your fists and stretching out your fingers. Let your hands fall into a natural position, with fingers slightly rounded. When presenting, be cautious not to point at the audience or show tension in your hands—they are often the first place where non-verbal stress is subconsciously revealed to the audience.

Various parts and portions of the body give cues to what the speaker truly thinks or believes, over and above the simple words spoken. The following outlines these various aspects, as well as the importance of non-verbal communication in team presentations.

Facial Expressions

Most of our non-verbal cues are drawn from facial expressions. The audience can see the movements of your eyes, mouth, and facial muscles and will evaluate your performance on these movements. While it is not always appropriate to smile at the audience (in fact, smiling can be perceived as incongruent with the topic being presented), an open facial expression where the face is relaxed and emotions can be easily detected is generally preferable to a closed, unexpressive face.

Many presenters identify eye contact with the audience as the most difficult aspect of the presentation. A time-honoured technique is to not look audience members directly in the eye, but to look at a central point on an audience member's forehead. The speaker may scan the room and select several audience members for application of this technique (useful in a large group), or may connect this way with every member of a small group.

The advantages of using this technique are two. Both the speaker and the audience benefit. The subject audience member perceives that the speaker is looking directly at him or her, and other audience members will feel that the speaker is connecting with the audience. The speaker is able to build this connection without being

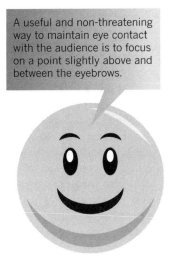

Figure 11.2 Maintaining Eye Contact

overly impacted by the facial expressions or non-verbal cues from the audience. Hold the engagement with an audience member for a moment or more, not so short that your eyes are in constant movement and not so long that the audience member becomes uncomfortable. The personal attention you give focuses the audience's attention and adds to their comprehension of the spoken words. With practice, you may become more comfortable with looking directly at audience members, but this technique remains useful for those situations where you are delivering a difficult message or sensitive information to an audience.

Practical tip. Do not roll your eyes in a presentation. While this may be a normal mannerism to show disbelief or disagreement (and practised by many teens on their parents), it looks freakish to the audience to see the lower whites of the presenter's eyes. Unless you are intentionally trying to scare the audience, break this mannerism when presenting.

Upper Body Movement

The movement of the hands, arms, and upper torso are second to facial expressions in giving non-verbal meaning to a presentation. It is important to note that gestures can be context-specific: this is critically important in a world growing increasingly smaller, where culturally based meanings can be very different from one social group to another. The same gesture may be interpreted as aggressive by one group, rude by another, and innocuous by a third. However, this should not be viewed as limiting presenters, requiring them to remove all gestures from their presentations. Instead, the speech becomes an opportunity to use non-verbal communication in a way that crosses cultures, a challenge to you, as the presenter, to advance the level of your performance.

The movement of the upper body is often the area where presenters' words subconsciously conflict with their gestures: they simply are not aware that their arm and hand movements are undermining their messages. Fortunately, once presenters realize gestures are an issue, improvements can be made rapidly in co-ordinating words and gestures. A few small changes, such as switching from a pointing finger to an open palm with the fingers touching, which presents a firmer gesture, can be made quite easily and will improve the congruency of the presentation.

Often, the presenter's hands give away his or her nervousness or create incongruence between what is being said and what the audience perceives. The presenter may be saying how pleased he is to be presenting, but nervously fidgeting with the hem on his jacket belies his words. The audience detects this discordance between gestures and words, and will (both subconsciously and consciously) begin to doubt the veracity of the presentation. Unfortunately, most presenters are not aware that their gestures are contrary to their words, and that these non-verbal cues are the first level of meaning for the audience.

Consider your normal 'hand stance' while you are presenting. One way to do this is to videotape a rehearsal of your presentation, or a perhaps simpler method is to present in front of a mirror while noting your gestures. You may also ask trustworthy colleagues if they have noted any repetitive or off-putting gestures in your presentations; but be prepared for the feedback you have requested.

If you find that you are unsure of what to do with your hands, keep them relaxed at your sides, or hold a pen or notes in one hand to give you a focus for your energy. Although some texts will recommend that you place your hands behind your back, this pseudo-military pose is aggressive and does not add character to the presentation (unless you are in the military and presenting to other military personnel, in which case this might be an acceptable, although limiting, stance). Others will recommend that you place one or both hands in your front pockets; however, this is not recommended because you remove your ability to gesture, your posture becomes rounded, and the tendency to play with objects in your pocket (like rattling coins) can be extremely distracting for the audience.

Generally, open gestures are perceived as more friendly and honest than closed gestures:

- the open palm facing out to the audience versus a closed fist;
- squared shoulders fully facing the audience versus rounding away from the audience;
- relaxed arms versus crossed arms;
- open arm gestures that begin at the shoulder and use the whole arm versus inhibited, small-action gestures moving out from the hands and wrists.

Guard against non-specific hand movements (fiddling with clothing or jewelry, fluttering hands, or jerky actions)—the human eye follows motion and these random movements will distract the audience's attention. In theatre, one way to upstage a fellow actor is to move while that actor is speaking. The audience cannot help but be distracted by the movement, and the speaking actor will lose

---

Box 11.2 Use of a Pen

Practise this: pick up a pen, lay the pen across your palm, and hold it without too much tension (do not hold it as though you will be using it to stab the audience). Read a selection from this text, and consider what hand/arm movements could be made to support the selection.

If you are speaking of measuring the extent of something, you may move your hands apart. If your words move around, defining or explaining an idea, you might end by putting your hands together.

Do not fiddle with the pen; use it as a tool to keep your hand relaxed and ready to make gestures that support your presentation.

her connection with the audience. Do not use gestures that distract the audience from your own presentation!

Scale the size of your gestures to the size of the room, while keeping the gestures natural. Try this exercise: place your arms at your sides, glued to your body. Gesture only with your hands, moving them from the wrist downward. Note how uncomfortable this is, both in feel and appearance. Now gesture only from the elbow down (keep the upper arms and shoulders frozen). Again, this gesture is awkward and unnatural, yet is one that many presenters assume when speaking. Now gesture with your whole arm from a relaxed shoulder. Feel how comfortable the gesture is. Gestures that emanate from the shoulder project better in any venue, small or large. Full-arm gestures look more relaxed, and make better use of the abilities of the presenter. Picture the difference in the size of the gesture you would make from a small to a large venue. The actual difference may be small, but held longer or expanded in a wider arc in the larger venue. Tailor your gestures to emphasize important points, express your emotions, and bring the audience into your presentation.

Gestures that match what you are saying are useful when presenting and add life to your presentation. However, too much gesturing is distracting. The human brain is wired to follow motion, and when motion does not correspond to the words being spoken, the non-verbal cues will take precedence and the audience will no longer be listening to what you are saying. Find a balance by rehearsing your presentation and becoming aware of any distracting habits you may have that will negatively impact the presentation. Your upper body movements reveal your true feelings, motives, and intents in the presentation: be certain they are congruent with your message.

Lower Body Movement

While nervousness is most often revealed in hand gestures, the movement of the presenter's feet and legs is the second most visible area for revealing tension and stress. The least powerful stance for a presenter is 'the stork', where the presenter shifts all weight to one leg and either actually picks up the other foot and waves it while speaking or balances on the toe of the foot. In either case, the presenter appears uncertain and tentative, the presentation loses verbal and non-verbal congruence, and the presenter is not believable.

Being seated for a presentation also requires that attention be given to the feet and legs. A draped table does not always fully hide the movement of feet beneath it, and legs that are crossed can appear awkward and uncomfortable. Even if the audience cannot see the presenter's feet or legs, the movement of the fabric is extremely distracting.

Whether standing or sitting while presenting, keep both feet firmly rooted to the floor, moving them only if the movement adds to the presentation. If standing, shift the weight from one foot to the other if you are doing an 'on the one hand' comparison, or step towards the audience if you are seeking to pull them into the presentation. If seated, keeping the feet on the floor is preferable to any shifting, leg-crossing, tapping, or fidgeting.

For many presenters, the 'safe zone' is the space immediately behind the lectern. They will grip the sides of the lectern as though they will fly off into space if they let go, and will remain rooted in that place. Increasingly, however, cordless microphones allow the presenter to move about the presentation space to give more successful presentations. An excited, animated performance may draw the audience in, with the kinetic display adding to the interest of the presentation. The speaker may show her control of her topic and the space by moving strongly from one area to another, and may use presentation materials situated around the presentation space as anchors for this movement. The audience remains engaged as they are locked to the motion of the presentation, and the presenter is able fully to use non-verbal communication to express her interest in the topic.

Practical tip. Use a critical eye to evaluate the performances of public speakers—consider a range of sources from politicians to televangelists. While one might initially expect that all politicians would be accomplished public speakers, many use a lectern as a crutch, standing rigidly behind it as a barrier between themselves and the audience. Many televangelists, on the other hand, are masters of using the space on the stage, and may even enter into the audience's space or pull audience members on the stage. Find new opportunities everyday to evaluate public speakers and the techniques they use to help to expand your own range of presentation techniques.

The power of moving towards something (a presentation screen, flip chart, or even the audience) is important to note: the audience's attention is immediately riveted on the object towards which the presenter moves. Use this technique if you are attempting to draw attention to a particular slide in a PowerPoint presentation or if you want to draw the audience in on a key point in your speech.

As with many of the techniques used in presenting, care must be taken to match your movement to the words you are saying. Random stalking around the stage or bursting into the audience will likely brand your presentation as odd rather than informative. The presenter who paces back and forth like a caged tiger will create a high level of discomfort in the audience, which is only useful if that is the presenter's intention. The presenter who trips over a projector cord will not be viewed as credible by the audience. A great deal of movement may also cause technical issues, for example, if the only way to advance the PowerPoint presentation is from the lectern, or if the speaker's movement is limited by a corded microphone. (For discussion of the use of visual aids in presentations, see Chapter 15.)

As with any presentation technique, you should practise moving about the space in advance of the presentation. Identify any issues well in advance of the presentation to ensure that your intended movement does not interfere with the words being spoken.

Team Communication

If you are presenting as a member of a presentation team, the non-verbal communication among team members is critical. Within the team, ensure that your non-verbal cues are supportive:

- Maintain an open and agreeable facial expression when other team members are speaking.
- Maintain a body stance turned slightly towards the presenter.
- Maintain stillness in gestures and movement while someone else is presenting. Be aware that the audience's eyes follow motion, and they will be distracted by movements unrelated to the speaker's words.

In some cases, appropriate physical contact between presenters can be used (a hand on a shoulder, for example) to show the connection between presenters.

Eye-rolling, a closed posture, and unnecessary distracting gestures all tell the audience that you do not agree with or support the other presenter. These latter techniques can be useful if you are in a debate with other presenters on the stage. However, great care must be taken not to go so far as to turn the audience against you, if they interpret your gestures as unnecessarily rude.

Overall Appearance

Is your overall appearance congruent with your message? Often, presenters will falsely believe that the power of their words will overshadow any incongruence between what is spoken and personal appearance; for example, that overalls are a suitable outfit for presenting in the boardroom; that a suit and tie are correct when presenting in tropical heat in a rain forest.

Two aspects must be considered:

1. being true to the presenter's own personal image and message;
2. being appropriately dressed for the venue and audience.

For the first, the presenter must consider how the message is being represented in his or her individual style. The audience expects a high level of congruence between the speaker's appearance and the message, but will give the presenter a great deal of latitude. The audience may have no preconceived notion of what the speaker should be like, only that the appearance and the topic must somehow tie together.

At an academic conference, for example, one will see great divergence in the style of dress of the presenters: the business suit and tie crowd; the rumpled corduroy blazer, unpressed khakis, black socks and sandals group; the 'alternatives' with piercings and vintage clothing; and the 'regulars' who fit in no defined style cohort other than non-descript and vaguely traditional. Multi-ethnic communities bring presenters in clothing styles relating to their cultural homelands. In short, there is no prescribed style of dress, and the audience is generally willing to allow the presenters a great deal of leeway.

This is not true for other presentation venues: a presentation to corporate bankers on Bay Street requires a highly conservative standard of dress, even on 'casual Friday'. Cultural mores may also dictate particular requirements. Depending on the nation and venue, presenters may be required to adopt local customs and standards to be permitted to present. At issue is the balance between personal style and the expectations of the audience for the presenter. While some situations are reasonably clear-cut (like presenting to the corporate bankers), others are less so.

Before presenting, consider whether or not your image works with the intent of your presentation. A too-casual outfit may undermine your credibility as a presenter while an outfit that is overly formal may look as though you consider yourself superior to the audience, do not understand them, or are more than a bit naïve and really don't have a clue. When in doubt, discuss appropriate attire with the presentation organizers or aim for a 'value-neutral' outfit that will not undermine your presentation. As an example, many male North American government officials doing business in tropical or equatorial nations will adopt khakis and a short-sleeved dress shirt as a uniform for doing business. While it does not speak to individuality, it is generally innocuous enough to generate acceptance in a wide variety of situations.

On occasion, a presenter will defend his personal choices with, 'It shouldn't matter what a person looks like if their message is important!'—which is true to a point. However, knowing the weight the audience places on the non-verbal aspects of a presentation, it makes sense to be sure that the audience is given every opportunity to find congruence between your appearance and your message. Do not undermine your own abilities as a presenter by making choices that do not fit either your personal style or the intended audience.

Improving Non-verbal Communication

When a speaker is comfortable presenting, he or she is much more likely to maintain a positive expression, open gestures, and a strong stance. Many aspects of non-verbal communication flow naturally from this comfort: if a presenter is enthusiastic, prepared, and knowledgeable, she likely will subconsciously reveal non-verbal cues that create a believable presentation. However, if a presenter is nervous or ill-prepared, non-verbal cues will distract from the verbal message and overpower the presentation. Fortunately, there are two simple solutions:

1. *Be prepared.* What works for Boy Scouts works for the presenter. Be prepared for any eventuality and this confidence will be reflected in the presentation.
2. *Be aware.* When rehearsing a presentation, observe whether any of the negative non-verbal communication actions shown in Box 11.3 are present, and actively work to replace them with the corresponding actions from the positive action column.

Box 11.3 Negative and Positive Non-verbal Communication

Negative	Positive
• Hiding behind the lectern	• Using the lectern as a supportive prop for the presentation
• Speaking to only a portion of the audience	• Speaking to all audience members (may require moving away from the lectern and about the stage)
• Posture rounded, chin down	• Posture 'ready', chin parallel to floor
• Standing on one leg	• Feet rooted to floor
• Turning from audience	• Fully facing audience
• Leaning away from audience	• Relaxed stance or leaning slightly forward
• No eye contact	• Direct eye contact
• Staring (more than culturally acceptable)	• Holding eye contact for an appropriate length of time
• Shifting eyes	• Holding eyes steady
• Eye-rolling	• Do not roll eyes upward
• No smile/frozen expression	• Open and changing expression to suit topic
• Crossed arms	• Arms at sides, relaxed
• Clenched fists	• Relaxed palms, fingers touching
• Hands in pockets	• Hands relaxed at sides
• Pointing	• Relaxed open palm with fingers touching
• Legs crossed	• Feet rooted to floor

With knowledge and practised skills, your verbal and non-verbal presentation aspects will become fully integrated.

Reading the Audience

Just as the audience is reading the non-verbal aspects of a presentation, the presenter can 'read' the audience as well. Depending on the size of the audience and the presenter's abilities to tailor the presentation, it can be useful to react to visual cues and be receptive to the needs of the audience.

For example, if the majority of audience members are starting to fidget and check their watches, it is likely that the presenter has reached the end of the presentation time and needs to put a finish on the speech (whether or not he has completed all the material). A series of non-verbal cues conclude many university classes: students will begin to pack up their materials, unplug laptops, put on jackets, and otherwise indicate (non-verbally) to the professor that class is finished. Without discussion, the professor will complete the lecture. The students and the professor both are aware of the intent and power of the non-verbal cues, and the students' actions are a culturally acceptable means of letting the professor know the class time is over.

In situations where the norms are not so clear-cut, the audience can still give very obvious cues to the presenter, including those listed in Box 11.4.

It is important to note that the same mannerisms do not always carry the same meaning for each audience member. As an example, many people naturally sit with crossed arms as they find this a comfortable posture. Again, cultural differences may exist, so be aware of the danger in reading too much into a particular pose or action.

More importantly, a speaker should be in tune with the audience: if the speaker is making a point and many people in the audience squint or rub their eyes, it is likely that the majority of people did not understand the speaker. It would be

Box 11.4 Cues from the Audience

Signal	Meaning
Fidgeting, watch-checking	Presentation time is complete (or it is lunch time?)
Smiling, eye contact	Interest in the presentation
Folded arms	Disbelief (or that may be a comfortable position for that person)
Head-scratching, rubbing side of nose, squinting	Do not understand point being made
Taking too many notes	May not understand point; can also reflect boredom
Fiddling with pens, objects	Boredom, not paying attention
Staring without blinking	Not listening
Head tilted to one side	Listening, engaged
Head nods	Falling asleep
Shuffling, fidgeting	Distracted

worthwhile, then, to review the point or restate it in a more easily understood manner. Be perceptive and you may be surprised at how many non-verbal cues can be received from the audience.

Effective Non-verbal Communication

The most effective presenters are those who are personally interested in what they are presenting, use gestures and movements that support what they are saying, have good eye contact, and, most importantly, maintain congruence between the verbal and non-verbal aspects of their presentation. The effective presenter understands that the audience's interpretation of non-verbal aspects of the presentation is many times more important than the spoken word; this is not to undermine the content of the presentation, but to emphasize that the presentation is more than words and phrases; it is movement, expression, and gesture as well.

Anyone with a command of language can stand in front of a group and read words from a page. A good presenter brings life to the presentation and increases the audience's understanding of the material.

Chapter Review

This chapter examined non-verbal communication—the signs that you give an audience as a speaker and the signals you receive from the audience. As with any aspect of presentation skills, if the presenter has any habits that are distracting or detract from the presentation experience for the audience, these can be corrected. A good presenter is aware of the importance of non-verbal communication and the impact it can have on a presentation.

Review Questions and Activities

1. What is the significance of non-verbal communication to a presentation?

2. Name three non-verbal cues that you have witnessed from an audience, either as a presenter or as an audience member. How effective was the presenter in reacting to these non-verbal cues?

3. Consider your own actions as a presenter. Are you giving non-verbal cues to the audience? Are they consistent with your message?

4. Spend one day observing non-verbal cues in every situation. What do you notice? What has surprised you?

5. Explain a technique that can be used for a speaker uncomfortable with making eye contact with the audience. Why is it effective?

Chapter Twelve

Creating a Great Presentation

You have been asked to make a presentation. This may be part of an academic requirement for a mark, an identified career responsibility, or because you have knowledge that is perceived to be of value to an audience. How do you begin to craft a presentation that provides useful information, meets specified requirements, and is interesting and possibly enjoyable for both the audience and the speaker? This chapter outlines a process for creating an effective presentation.

Purpose

Have you ever attended a lecture, allegedly on a defined topic, and walked out wondering what happened? A *clearly defined purpose* is an agreement between a speaker and an audience. The speaker will identify the purpose of the presentation, and the audience will listen and observe to see if the speaker meets this objective. While the presentation may cover much more than a single, defined purpose, the success of the presentation requires the clearly stated achievement of this part of a presentation.

An undefined purpose leaves an audience searching for meaning and structure in the presentation. What was the presentation about? Why was the audience left to riddle out the purpose? In some situations, the purpose of the presentation is implicit—at a press conference, the media are aware of the purpose of the event because it was clearly outlined in the press release. However, for an academic lecture on 'The Greening of Fort McMurray, Alberta', a stated purpose is required to let the audience know what the speaker intends to cover on this formidable topic.

Even worse than an undefined purpose is a defined purpose that is not achieved and leads the audience to believe that the speaker's understanding of the subject is incomplete. Do not overlook the importance of clearly understanding the purpose of the presentation both for the speaker and for the audience.

Ask yourself the following questions as you define the purpose of your presentation:

- What is the point of this presentation?
- Can I summarize it in one short sentence?
- To whom am I presenting? Do they have a prior understanding of the topic, or am I presenting new information?
- How can I best present this information? What format should I use (see below)?
- What do I hope to achieve from this presentation? What does the audience want from it? Are my goals congruent with those of the audience?

Without a clearly defined purpose, it can be difficult to craft an effective presentation. Work to discover what it is your presentation is about and the structure and shape of the presentation will quickly follow.

Practical tip. It is important to recognize that presentations are individual events, measured against the 'personal bests' of your previous presentations. The secondary purpose of any presentation is to retain those aspects of past presentations that have been effective, and improve on aspects that need additional work. Do not let past presentations limit your current performance—consciously decide how you want the presentation to unfold, and allow yourself to be successful.

Box 12.1 Template for Evaluating a Presentation

Title of presentation:

Venue (add positive and negative comments on venue):

Presentation date:

Time:

Duration (how long was the presentation?): Required: Actual:

Internal issues (presenter's state of mind):

External issues (noise, sound system):

Purpose of the presentation:

1. What went well in the presentation?
2. Where was I better than in my last presentation?
3. What would I do differently next time?
4. How did the audience react?

Know Your Audience

Are you presenting to a group of learned professionals with a high level of know-ledge on your topic? If so, be sure that your data are current and references are cited. Are you presenting to a general interest audience with limited information but a high level of interest in your topic? Avoid technical jargon or advanced lan-guage that will be unfamiliar to the group. An effective presentation considers the audience and is tailored to their needs.

Research your audience before the presentation. Ask the organizers who is expected to attend, how many invitations have been issued, whether the size of the audience is known in advance (if tickets or pre-registration is required), and if the event is open to the public. Generally, for most university presenta-tions, the audience is known to the student presenter: it is likely your peers and the professor for that particular course. However, if you are presenting at a conference or at a special event on campus, the size and composition of the audience may be undetermined. Career presentations may also have less well-defined audiences.

The most uncomfortable position is thinking you will be facing an audience of like-minded individuals and finding that the crowd is much larger and more hos-tile than anticipated (as is sometimes the case for meetings open to the public that do not require pre-registration). In this instance, you must rely on your learned presentation techniques, and be ready to go 'off script' and adjust the presentation to the needs of the audience. You may have planned and rehearsed a 30-minute educational presentation, but the non-verbal cues and the tone from the audience suggest that they have no patience for it. A gracious and prepared speaker will be willing to shift to a question-and-answer format and work in partnership with the audience (see Chapter 13).

Venue and Technology

The venue and available technology should be considered early in the presenta-tion development process. If you are presenting to a handful of people in a small room with only a flip chart, your presentation will take a different shape than if you are presenting in a large theatre with hundreds of people in attendance and the latest in presentation technology. While your thesis and content may not dif-fer so much, the way you present should be tailored to maximize your use of the presentation space.

As a prepared speaker, you will have investigated the venue prior to your presentation (or, if that is not possible, you will have at least spoken to someone who knows the venue well). In a small space with a small audience, it may be more comfortable for both you and those in attendance if you remain seated and the presentation takes on a more conversational tone. Your words and non-verbal communication will be as well researched and prepared as they would be for a more formal delivery; the only difference is in your ability to tailor the presentation to the venue.

In a large venue, is AV help available? Where will you be in relation to the screen (if you are using visual aids, which would be recommended in this large space)? How does one move from one slide to the next in a presentation—a hand-held device or only from a button on the lectern? Is the microphone fixed or cordless, and how sensitive is it? Asking a series of questions before the presentation and rehearsing in the presentation space are two ways of ensuring the success of your presentation. As noted previously, credibility is lost if you appear unable to manage the technology in the venue.

Creating and Designing the Presentation

Many textbooks on presentation skills will tell you to begin your presentation with a humorous anecdote, but *not this one*. More often than not, the audience is left wondering at the relevance of the joke, detracting their attention from the topic at hand. A better opening tells the audience what you are going to do— what is the purpose of the presentation, and what do you hope they get out of it. A story or anecdote may follow as a means of explaining the topic ('this reminds me of the time that . . .'), and then the audience will have the advantage of clearly understanding what they are supposed to be hearing from you.

Your opening sentence should be powerful, concise, and clear: the audience will know where they are going and how they are going to get there. A classic speech format is to tell the audience in your opening sentence what you are going to be speaking about, then proceed to speak about this, and then finish by telling them what you told them. This three-part structure works to reinforce your presentation message and gives the audience additional opportunities to understand and remember your key points.

The danger is in creating a presentation that feels over-structured to the audience. Do not say, 'Now I am going to tell you what I told you.' Instead, let the audience in on the purpose of your presentation, give the audience a 'heads up'

Box 12.2 'Heads Up'

An effective technique for making sure the audience is ready to hear a key point is to give them a 'heads up'—a statement that precedes your key point and lets them know that what follows is important. 'Heads ups' include:

- The most important thing I am going to tell you is this . . .
- I'm going to let you in on a secret . . .
- If you do not remember anything else from today, remember this . . .
- The *true* story is . . .

For even greater impact, follow the 'heads up' with a pause, not so long that you lose the audience but long enough to allow them sufficient time to get ready for your key point.

when restating key points (Box 12.2), and be sure that they have a third opportunity to rehear and understand your thesis in the conclusion.

To start developing your presentation, begin with this three-part 'tell' structure. The purpose is the 'what you are going to tell them', and your introduction is critically important. You have a few brief moments to grab the audience's attention and establish your credibility as a speaker. Plan to create an impact: what would you be interested in if you were sitting in the audience? What would grab you? A powerful and well-rehearsed opening allows the speaker and the audience to relax into the presentation.

Next, add in the second level of information: the data and evidence that supports your thesis (the 'tell them'). Again, remember the *rule of three*—the audience is unlikely to remember more than three key points. Therefore, you do not need myriad statistics and case studies to support your thesis. A few carefully selected pieces of support material are better than reams of evidence that is overwhelming and quickly forgotten. The added benefit to this simplified presentation is that it is easier to deliver for the presenter, and the presenter is much less likely to go off track or lose her place in the presentation if she is working around just a few key points.

Once the presentation begins to take some shape, add colour to the presentation with stories, anecdotes, or quotations (Box 12.3). Often, the best stories are your own personal experiences, particularly if they are humorous or show the fallibility of the presenter.

Quotations are often well received and can add a note of integrity and reliability to your presentation. The only caveat is to be certain that it relates to your thesis; a random story quotation might be interesting or amusing, but an opportunity has been lost if it does not add to the presentation.

Practical tip. Be careful if you are poking fun at a member of the audience. What works for stand-up comedians may not work at an academic conference.

Box 12.3 Add Colour to a Presentation

- Stories
- Anecdotes
- Video clips
- Cartoons
- Graphics/charts

- Questions
- Quotations
- Pictures
- Polls of the audience
- Statistics

- Exercises to involve the audience (meet your neighbour, request an assistant, ask a question of the audience)

Using graphics (photos, charts, and diagrams) as part of your presentation is generally expected by most audiences. We live in a highly visual society, where presentations are increasingly taking cues from the media and shifting towards fast-moving, image-packed formats. A note of caution, however: do not let graphics take over your presentation, unless you are presenting on the use of graphics. Remember that your key objective is to have the audience retain a few key points and understand your message. If your presentation also happens to be visually stunning, so be it. Just do not let the ability to create a highly visual presentation detract from your thesis.

Let your presentation have an ebb and flow by using all of the elements of speaking considered in Chapter 10: let the pitch of your voice and your volume rise and fall with the importance of words being spoken. Speak more quickly if you are telling a particularly exciting anecdote, and more slowly as you specify the main point of your presentation. If you are comfortable, move about the presentation space and draw the audience into your presentation. Do not, however, let your energy or enthusiasm wane. If the audience senses that you have lost interest in what you are presenting, you can be certain that they will lose interest as well.

The third and final part of the presentation is restating 'what you told them' in a high-impact conclusion. When developing your presentation, plan to finish. This may seem obvious, but so many presentations end with 'Um, I guess that's all there is' or another form of an awkward pause.

What do you want the audience to remember? You have one final opportunity to deliver a key message. Ideally, your final words will resonate with the audience for years to come. Deliver a short, succinct statement that powerfully sums up your message, and consider preceding it with one of the 'heads up' statements. Use presentation techniques such as a louder volume, a slightly lower pitch for a serious statement or higher to illustrate excitement, an added powerful emphasis on the key words, and a down inflection at the end of the sentence to signify that this is the end of the presentation. The audience will appreciate being properly cued on the conclusion to the presentation so they can be prepared to listen to the final statements.

Formats

Your presentation may take any number or combination of formats, and the format should match you as a presenter, your topic, and your audience. If possible, consider developing your presentation in more than one format—that way, if you need to tailor your presentation to a new situation or audience, you will be prepared and ready to shift into a more effective format for that circumstance.

Lecture

The most common presentation format is the lecture, often used as the default presentation format without any consideration of other formats that may be more effective in conveying information to the audience. The lecture holds some advantages, however: it is an expected format, and the audience is allowed to take on the comfortable role of listening. In addition, a lecture is an efficient and highly controlled means of presenting as the speaker determines the content and the audience receives the information.

This format can be limiting because the level of engagement required of the audience is low, and low retention of information may be the result. To increase the limited role of the audience, the lecturer may add a question-and-answer session or other formats noted below to increase audience involvement.

Demonstration

A demonstration shows the audience how something works. A demonstration can be a powerful technique for increasing understanding, and audience involvement can be increased with drawing in audience members as recruits to assist in the demonstration. Even for the audience members not chosen, their familiarity or identification with the recruit will cause them to retain a higher level of interest in the presentation. The speaker's control of the situation is lessened, but the audience will be more focused and will retain more information if the demonstration is successful, interesting, and relevant.

Panel Discussion

You may be part of a panel or group of presenters, a format that is used often at conferences or academic gatherings. Each presenter is given a short time frame to present on her or his topic or area of expertise, then the presenters either discuss the topic among themselves (with the audience as passive viewers) or the discussion is opened to the audience for questions to the panel.

While the panel discussion has the advantage of combining speakers and viewpoints on a topic, a presenter should be aware of the leanings of other panel members and be prepared for real conflicts or fabricated ones (intended to add drama to the panel discussion). You should also be aware that other presenters may try to dominate the discussions and the question-and-answer sessions. If possible, confirm with the organizers that each presenter will be given an allotted, uninterrupted time to present, and request that the question-and-answer session be managed by an external facilitator.

Audience Discussion

In this format the presenter raises an issue, then opens the floor to the audience. Audience members direct their comments back to the presenter, who may respond to individual comments or go on to the next speaker. The questions may be stacked in the audience (that is, the presenter or organizers have assigned individuals in the audience to ask predetermined questions—and yes, this does happen) or the presenter may allow for a free-flow exchange of information and then provide summary comments. While this format can be effective with an audience with a high level of knowledge on the topic, more often than not it diverges into discussions far removed from the purpose of the event.

Question-and-Answer Session

In this situation the presenter does not provide information to the audience without being asked a specific question. Often, a lecture-style presentation will end with opportunities for questions and answers, providing a combination of formats, or this format may be used alone in a press conference or other interactions with the media. The various problems and possibilities of question-and-answer formats are discussed in Chapter 13.

No single format is 'correct', although the lecture followed by questions and answers is the most common. If you have the opportunity to suggest a different format for your presentation and are willing to try something new, consider one of the above formats or combine different approaches to customize the presentation to the audience.

Prepare, Prepare, Prepare

Just as *location, location, location* is the mantra for real estate, *prepare, prepare, prepare* is the mantra for the effective presenter. A profound difference in tension is felt when you are prepared but slightly anxious about a presentation, as opposed to when you know you are woefully unprepared (and soon the audience will know it, too).

Prepare presentation materials and test them on a trusted colleague. Are the slides in a logical format? Do the charts and graphs make sense? Are they legible? What may seem obvious to you (because you know so much about the topic) may not be understood by the audience. If the presentation materials do not help the audience to relate better to your key points, rework them until they do. Many presentations contain random bits of information that are more distracting than helpful. Be certain that every slide, every overhead, and every word written on a flip chart is useful and supports your thesis.

Be prepared to deal with different presentation technologies, as well. The laptop and projector system at the presentation venue may not be the same as the ones you are familiar with, but the success of the presentation still rests with the presenter. Know which function key connects the projector and laptop for different brands of laptops (often, the 'Function+F4' combination works, but not for every make and model). Learn how to adjust the *internal* volume on a variety of laptops as this will impact the maximum volume of attached speakers. Be adaptable when faced with new technology and you will likely find that solutions can be easily found.

If you are quoting statistics in your presentation, are you working with the latest available information? The credibility of the presentation is questionable if you are working with dated data or quoting references more than a few years old (unless you are giving a historic overview of an issue). New census information is available every five years (and special reports are published frequently); many organizations produce annual reports. As the expert on your presentation topic, it is your responsibility to remain current in your references.

Practical tip. Are you certain your quoted data are correct? If you have collected the information through primary research, be sure to document your research techniques and methodology and be prepared to defend the thoroughness of your research. If you are using secondary data sources, only use data from reliable organizations, such as government census materials for population information. Not every website should be considered a good data source: triangulate (or check) your information to be certain that any referenced statistic is correct. You can be certain that someone in the audience knows something about the topic and would be thrilled to point out any errors in your research or presented data.

Practise, Practise, Practise

Our abilities as public speakers improve with practice. It is interesting that people will put a great deal of effort into developing a desired skill (getting to the next level on a video game) and may even spend money on skill development (getting private lessons at a ski hill), but they will walk into a presentation without ever having rehearsed their speech or thought about how they might respond to questions from the audience.

Practise your presentation. This means doing more than just thinking through what you might say: actually say it! Memorize your opening and closing sentences. Practise your stance and gestures. Map out places where greater emphasis, a higher pitch, louder volume, or faster pace could be used to add life and excitement to the presentation. Brainstorm contingencies. What will you do if something goes off track? What if the projector fails? What if you lose your place? Work to have a Plan B for any contingency.

Consider all the elements of speaking and presentation techniques, and look for ways to work them into your presentation. For example, is there a place where a 'heads up' and pause combination would add to the delivery of your message? Would moving towards the audience while maintaining eye contact with an audience member help to create new focus for your presentation? Let yourself add one new presentation technique to each new presentation, and attempt to vary one of the elements of your speech. Be conscious in improving presentation skills and you *will* be a better presenter.

Delivering the Presentation

You will recall that the audience likely will retain only three main points from your presentation, along with their impression of your non-verbal communication cues and the tone of your presentation. This is good news for a presenter!

As a well-rehearsed presenter, you are interested in and enthusiastic about your topic. Therefore, you will find it is easy to have your words be congruent with your non-verbal communication and let this energy flow through your facial expressions and gestures. Also, as a well-rehearsed presenter, you have created a good opening and closing, have key points identified, and are ready for any contingencies that might present themselves.

Only a few outstanding issues need to be addressed in delivering a great presentation.

Time Frame

How long do you have to speak? While the topic may be the same in a 5-minute or 20-minute speaking engagement, the contents and level of detail will differ markedly. Even when you are aware of the allotted time frame and prepare in advance, too often you will find yourself rushing the presentation at the conclusion to fit in required information because time has run short.

The reasons for this may be beyond your control—earlier speakers went into 'overtime' and cut into the time allotted for your presentation, or perhaps technical difficulties delayed the event. Conversely, the reasons may be entirely of your own making and attributable to the success of the presentation—perhaps you went 'off script' to tell an interesting or relevant story, or time was lost to audience laughter or to interaction. An effective presentation is one that can be tailored to fit an allotted, and possibly variable, time frame, and an effective presenter knows when to move along if time is progressing too quickly.

Agenda

How many speakers are on the agenda? Are you the keynote (main) presenter or the opening act for another speaker? Where are you on the agenda? If possible, avoid being placed at 'low energy' points in the day (right after lunch, 4:00 pm after a day of speeches, on the last day of a conference) or at times when the audience may be distracted (right before a big event like a keynote speaker or major social gathering).

Use of Notes and Slides

The use of speaking notes is highly individualized. Some people prefer to have their entire speech in front of them, although they may seldom refer to their notes. Others work from cards and key points, and still others can present from a scrap of paper with a few words on it. Often, as presenters become more experienced, they are able to work with fewer notes and the presentation flows more naturally.

Box 12.4 Remember . . .

- Have a strong purpose.
- Know why the presentation is important to both you and the audience.
- Structure your presentation with signposts and enough repetition so the audience can follow along.
- Practise—especially the opening and closing.
- Recognize the limits of a human's ability to retain information and work to ensure your key messages are remembered.
- The audience wants you to succeed—do so!

If you do keep notes in front of you, never read your presentation to the audience. Nothing more swiftly kills the spark of energy in a presentation than having it recited aloud, word for word. In addition, never, *never* read the content of PowerPoint slides to the audience. Unless you are presenting to a group of preschool children, it is likely that most of the audience can read (and if you are presenting to non-readers, why are you using text on the slides?). It is annoying to have a presenter read slides verbatim when the audience members are perfectly capable of reading on their own. Ideally, as the presenter, you will not look at the PowerPoint screen except to ensure that you have moved to the next slide.

Be sure you know the order of your slides. A presenter should not be surprised by his own presentation. The speaker loses credibility when he seems shocked by the content of the next slide ('Oh, I didn't know that was in here!'). If you cannot follow your own presentation, the audience will be lost, too.

If you are presenting from a lectern, hold your notes to the top of the lectern. Keep track of where you are in your notes by running a finger or thumb along the edge of the paper. This way, your eyes do not have to scan far down the lectern and your chin can stay parallel to the floor, thereby largely maintaining eye contact with the audience.

Customizing Your Presentation

Whenever possible, add points to your presentation that will be relevant to that particular audience. A comment about the city or venue, a special thank you to the organizer of the presentation, or even a remark about the weather can help to draw a connection between you and the audience.

Let Go of Self-imposed Limits

Right now, decide that you will be an effective presenter in all future presentations. Presenting is a learned skill, and hanging on to negative beliefs about your presentation skills serves no useful purpose. Figure out your strengths as a presenter and use every technique and element of speech available to craft a great presentation.

Let your enthusiasm and knowledge shine through your presentation, and know that the audience wants to see an effective presentation that gives them new information. Use the following presentation checklist in Box 12.5 (photocopy and fill out in advance of each presentation) to confirm that you are ready (and excited!) about presenting.

Box 12.5 Presentation Checklist

- Know the purpose of your presentation (stated in one short sentence).
- Research the audience.
- Research the venue.
- Clear and powerful opening statement.
- Clear and powerful closing statement.
- Key points are established and clearly stated.
- Colour added to the presentation (only relevant supporting materials).
- Conclusion restates key points/thesis.
- Elements of speaking reviewed and incorporated into the presentation.
- Presentation has been practised a minimum of three times (full run through).
- Non-verbal communication has been reviewed and is congruent with the words.
- Eye contact—plan for establishing eye contact has been completed.
- Breathing and relaxation techniques practised and ready to be used if needed.

Just as anyone can learn to be a better skater or golfer by practising the skills necessary and anyone can add to her bench press by actually going to the gym, anyone can improve presentation skills. Deciding to spend one hour per week on presentation skills is an investment that will provide enormous returns in any career or venture.

Chapter Review

This chapter presented methods for developing an effective presentation. A good presenter uses every presentation as an opportunity to increase his or her personal skills. Each presentation is evaluated, areas for improvement are identified, and each presentation builds on the previous success.

Review Questions and Activities

1. Most presentations benefit from 'colour'—the addition of stories or anecdotes that give life to the presentation. Consider your last presentation. Is there a colour element that could have been added to improve the presentation?

2. Develop a method for evaluating your presentations. Use the form provided or create your own, and decide whether a paper or computer format would be more effective (the one you would be more likely to use). Consider your last presentation, and fill out the evaluation form. Do this for each presentation, moving forward to develop new presentation skills.

3. This chapter outlined several 'heads up' statements that can be used to focus the attention of the audience. What are they?

4. Most presentations follow a lecture format. Review the other formats for presentations. Which one will you use next?

5. List the qualities of a good presentation. What aspects of your own presentation style would you like to amend? How do you intend to make these changes?

Chapter Thirteen

Engaging the Audience and Developing Your Own Style

Most presentations end with the opportunity for the audience to ask questions of the presenter. Ideally, the presenter's level of knowledge and preparedness is high, and he or she should be able to address virtually any question with ease. The simple truth is that most questions asked will have little to do with the presentation and a great deal more to do with the person asking the question. The presenter must be prepared to respond to 'non-questions' as though they are legitimate requests for additional information or clarification on the presentation. The following outlines several question formats, differentiating actual questions from non-questions.

But first, two warnings.

1. *Do not* ask the questioner 'Did that make sense?' or 'Did that answer your question?' You are setting up a one-on-one relationship with the questioner that will be difficult to break out of, and the rest of the audience will be excluded. In addition, the questioner is almost obligated to think of another question since you have turned the audience's attention back to that person.
2. *Do not* get caught in a debate (unless you are debating, but that is for another textbook). If the back-and-forth exchange with one audience member extends beyond two volleys, use the technique outlined below to extricate yourself from this situation. Audience members will first mentally and then physically leave a presentation that appears to have degenerated into an exchange between two people.

Kinds of Questions

Information Questions

This is the kind of question most presenters expect to receive after a presentation. An audience member will ask for clarification on a key point ('Could you explain the theory of relativity again?') or for more information on a key point

('What are the long-term plans for dealing with the tailings ponds from the oil sands production?').

As the presenter, your knowledge of your topic must be sufficient to answer any informational question. Review your presentation from the audience's perspective. What kinds of questions would you ask? Is information missing? Should your presentation be modified to ensure the audience understands this critical information? Knowing your audience will help in addressing their information needs.

On occasion, you may receive a question that could be considered less than stellar, or one that you feel you have clearly covered in your presentation. Do not react negatively to the audience member. Simply answer the question and then move on. There is nothing to be gained from insulting an audience member and the audience will likely turn on you if they perceive that you are rude or arrogant.

I Know More Than You Do

Among the kinds of questions that presenters receive, the 'I know more than you do' is the most frequent type. The audience member may be an expert on your presentation topic, either in reality or only perceptually. In addition, the individual may be trying to attract attention from other audience members, as is often true in work situations when senior staff members are present. The questioner often does not ask a question, but makes a statement that shows how much she knows or speaks instead about his own project.

As a presenter, the difficulty may be in responding to a question that is not a question. A useful response is 'That's interesting . . . and it reminds me of (story on own topic).' The audience is pulled back to your presentation and you are given an additional opportunity to highlight your own research or thesis.

The Challenger

A more aggressive form of the 'I know more than you do' question is *the challenger*. The challenger, through non-verbal cues and spoken words, will attempt to undermine your credibility. One may ask, 'How do you know that?' or another, 'What is your reference for that statement?' As a presenter, you must fully understand your topic and have at the ready key references that you may provide to the audience. Be certain of the source of statistics and know that they are from a legitimate organization. Be ready to defend your presentation—not antagonistically, but because you are informed and able.

The challenger may also ask an impossible-to-answer question. For example, 'Are you *still* trying to destroy the property rights of all citizens in Suburbia?' If the presenter says no, it means that the presenter formerly destroyed property rights, but has moved on. If the presenter answers yes, then the audience realizes what a terrible person the presenter is. A useful response is to restate your thesis or purpose: 'As I noted before . . .', then re-explain a key point in your presentation. Move on to the next question from a different audience member.

The Confused

Virtually every presentation with a larger audience will contain a question from *the confused*. These questions have nothing to do with the presentation and no relationship to the topic being discussed. As a presenter, the only response is to say, 'If I relate your question back to my presentation (use opportunity to restate another key point).' The audience will know that the question is off-topic and will not expect a direct response.

The Trapper

If a questioner begins with the statement, 'So what you are saying is . . .' be very careful with your response. Take a moment to observe the speaker's non-verbal cues to see if there may be some less-than-honourable intent to the question. Be certain, if you intend to express agreement with the questioner, that the interpretation of your presentation has been accurate. If the questioner is not parroting your presentation, restate your thesis. Begin your response with 'What I am saying is . . .' and use your own words, not those of the questioner.

The Sharer

On many occasions, the questioner is not interested in asking a question, but wants to use the opportunity to *share a personal story*. In an age where boundaries on personal information appear to be fairly porous, this format is becoming increasingly prevalent.

If it appears that the questioner is not getting to any relevant point, it is reasonable to act as an advocate for the audience (who are now wondering how long this is going to go on) and say, 'I see where you are going. That feeds right back to (redirect to key point).' If the person persists, it is important for you to take an active role in salvaging the question-and-answer session. Tell the person, 'I think you have a lot to say. I'd be interested in hearing the rest of this after the presentation.' Then go on to the next question.

What If . . .

Preschoolers often play the 'what if' game as they learn about cause-and-effect relationships. As adults, the 'what if' question may not be so innocent. Do not allow yourself to be drawn into speculating on hypothetical situations as the lack of scientific data or an agreed-on conclusion will open your response to an audience debate. State, 'While that is an interesting question, the research shows that . . .' and, once again, pull your response back to your presentation. Overall, the point is that as the presenter, you have the right to bring the questions back to your presentation. The audience will appreciate your attention to their questions and your management of the question-and-answer period.

It may be that you are provided with a moderator to manage the question-and-answer session. While this may at first seem to be helpful, the presenter loses the ability to select the person asking the question. If possible, manage the question-and-answer session yourself, as it is likely that the moderator has not been as observant, and moderators have no personal stake in the presentation. As the presenter, you have had an opportunity to carefully observe the audience and determine:

- who has been paying rapt attention;
- the non-verbal cues provided by audience members;
- whether or not you want to answer particular questions.

Practical tip. Do not lose the ability to select the questioner. You can then respond directly to those individuals who appear to be most interested and portray the most positive non-verbal communication. You then retain the right to extend the opportunity to ask a question to those who appear either less interested or hostile. The choice and the control remain with you as the presenter.

Final Tips on Engaging the Audience

- When responding to questions, keep your answers short. While the question-and-answer period is a good opportunity to restate your key points, the audience is not interested in a full review. A 10-second response is a good response.
- If possible, plant one or two questions with colleagues in the audience. This is not to make you look good by lobbing easy questions to you, but to get the ball rolling in the event that the question period opens with a painful silence.
- Choose the questions from different points in the room—while the audience may not consciously observe that you have selected questioners from across the venue, they will notice that it seems as though everyone in the room has had an equal opportunity to ask a question and that everyone is interested and involved. It can be difficult not to favour the front of the venue, as often the audience members most interested in your topic will sit in the first few rows. If the lighting in the venue is sufficient and you can see to the back, recognize one or more questioners from the last few rows.
- Use innocuous identification techniques to identify the person you have selected, such as 'the individual in the red sweater' or 'the third person from the end of the row' if the venue is very crowded and there are many people prepared to ask a question. Be careful in the use of gender typing, particularly if the lighting in the venue is poor. Never use ethnic identifiers, and be extremely cautious of labels that might be viewed as demeaning by some members of the audience; for example, the term 'girl' could be viewed as offensive when applied to a 40-something executive.

- When asked a question, give the audience member your full attention. Do not shuffle papers or drink water while someone is talking. Be careful of your non-verbal communication: you may be exhausted after your presentation, your mind racing with things said and not said, but if you have agreed to entertain questions, then the audience member has a right to be heard courteously.
- Restating the question from the audience can be problematic: if the entire audience cannot hear the questioner, audience members may call out to you to repeat the question. This is not harmful for clear informational questions, but can be if you are required to repeat statements from 'the trapper' or 'the challenger'. If it can be avoided, do not fully restate the question. Paraphrase the core of the question and bring the statement back to your presentation.
- Do not speculate on a response. If you do not know the answer to an informational question, say so. It is likely that there is someone in the venue with correct information who would be pleased to point out your speculative error.
- Be cautious about saying, 'That's an interesting question' to some audience members and not others. The audience member not complimented may feel slighted. Either do not use an evaluative statement or develop a repertoire including 'interesting', 'good question', and 'thank you for that'.
- On occasion, use, 'I would be interested in speaking to you after the presentation' for questioners with a great deal to say or for those interested in engaging in debate. However, do not say this more than once or twice for each question session or it will appear insincere. No one wants to talk to the entire audience after a presentation!
- If an audience member is extremely hostile and attempting to control the question session, retain your composure. Comment, 'You seem to have very strong feelings on this. Could we open the floor to more questions and see how others feel?' Or: 'Perhaps others have the same or a different viewpoint; would anyone who has not commented like to share his or her thoughts?' It is difficult for the hostile audience member to object to involving the rest of the audience; the audience will appreciate being involved back in the presentation.

A presentation has two roles: one for the presenter and one for the audience. Your role has been reviewed in this chapter, but the audience has responsibilities as well:

- to listen to the presentation;
- to respond with questions or comments;
- to be civil (although they do not have to agree with you).

It is reasonable to expect the audience will fulfill these responsibilities, and in most cases they will do so without any prompting from you, the presenter. However, sometimes the audience needs to have the presenter define their responsibilities:

- If there is too much chatter in the audience, it is reasonable for the presenter to pause and wait for the audience to return to quiet. A powerful response (but one that can be intimidating to audience members, so use it wisely) is to

pause and step towards the audience members who are causing the distur-
bance. Do not speak, but establish eye contact with the offenders, breaking it
only when the disruptive behaviour stops. Then, step back into the presentation space and continue with the presentation. The rest of the audience will
appreciate that the presentation has been pulled back into focus.

- If audience members cross-talk during the question-and-answer session, the
presenter can request that questions return to 'one at a time' so everyone in
the audience can hear the questions and responses. A statement on the value
of the questions to the entire audience will pull the audience back together.
Again, pause and let the audience regain its composure.

- If the audience starts to digress in a less-than-civil direction, the presenter
can redirect the questions to 'friendlies' in the audience—'Let's see if everyone feels that way.' Generally, the audience will regain a point of balance and
allow the question period to continue.

Is There a 'Right' Way to Present?

You have enormous control as the presenter. Your task is to present information
in your most compelling format, not to engage in debate or subject yourself to
negative comments. Retain control of the presentation and format and you will
find that individual audience members will meet their responsibilities, resulting
in a successful presentation for both you and the audience. At the same time,
you must bear in mind that there is no single 'right' way to present. Formats, as
we know, vary, as do our personalities. What works for one person might not
for another. Nonetheless, a presentation cannot be viewed as successful unless it
meets two important criteria:

1. Your presentation was *understood* by the audience. The most important measure of your success is that the audience understands your message. While
they may not always agree or act on your message, it is critical that the audience be able to get the point of your presentation.
2. Your presentation was *effective*—defined as credible, informational, and
enjoyable.

An approach that works for one presenter may not work for any other, and it
is important to give some thought to your personal style as you develop as a presenter. Part of the enjoyment for the audience comes from witnessing new, captivating presentation techniques and styles that are congruent with the presenter's
personality and subject matter. Some speakers use humour, others use facts, still
others employ props or visual aids: whichever format and style are chosen, the
most credible presentations are ones that work with the abilities of the presenter.
In other words, as a public speaker, you need to know yourself well and then use
your strengths to best advantage. But, always, be yourself; don't try to be what you
think an audience will want if that is contrary to who you are. Developing your
own style, therefore, involves knowing yourself and building on that.

Developing Your Personal Presentation Style

You may not consider yourself to be a natural presenter, and although you spend consistent time honing your presentation skills, you still would not rate presenting among your most enjoyable experiences. This characteristic could be made part of your presentation style; even if you are nervous, the audience will be receptive to a presenter who is sincere and provides useful information. Your nervous energy can be used to create a believable and effective presentation.

How can you develop your personal presentation style? Consider the most effective presentations you have given. Are there techniques, formats, or styles that work for you as a presenter? Consider the following:

- Are you a storyteller? Most presentations benefit from the telling of a personal story or the anecdote of another, as humans learn through examples and the audience is familiar with this teaching technique. Most speakers can effectively incorporate this technique into a presentation.
- Some people tell jokes easily; for others, jokes sound forced or contrived. Only tell a joke if you are certain it is funny and appropriate to the venue and if you have a good sense of timing.
- Do you move around the presentation area? Some speakers prefer to remain rooted behind the lectern, while others travel across the stage area or even enter into the audience. You may find that some movement helps to reduce nervousness (just be sure the microphone clipped to your lapel is wireless if you plan to be mobile).
- Are you likely to use props? Only use props you are familiar with and can comfortably use. For example, do not use a slide advancer if you tend to be nervous and will rocket through the slides inadvertently. Test the prop in advance and be prepared to work without it, if necessary.
  - One professional speaker who does more than 100 speeches a year learned quickly that she should not use a laser pointer. As she tends to use a lot of gesture, she found that she was pointing the laser at the audience, causing people to duck and squirm to avoid having the laser shine in their eyes. Instead, she keeps a collapsible pointer in her hand and opens it when she wants to emphasize a slide graphic.
  - An English professor I know ostentatiously dons a pair of heavy, horn-rimmed glasses on the day he teaches the literary use of irony. (They do not contain lenses, so he can see perfectly well wearing them.) The professor then struggles in his use of the overhead projector while he complains about the 'advanced technology' of the projector. The irony, of course, lies in his affectation of the wise-old-owl glasses when he is not smart enough to run a simple apparatus.
  - A student presenting on allowing community gardens within a municipality brought a selection of plants to the presentation and lined them up on the table in front of the lectern. The plants were highly fragrant and most of the students commented positively on how the plants added to the presentation

(except one student highly sensitive to odours). After the presentation, the student handed the plants out to class members who agreed to plant them in their own yards or within the university's community garden.

– A hydrologist attempting to encourage local governments to improve water quality brought a selection of samples obtained from a variety of sources in the area. At a particular point in the presentation, he put on a lab coat and unveiled a series of test tubes on the table in front of him. As chemicals were introduced into the test tubes, the contents changed colour, indicating contamination. The use of props like a lab coat and test tubes gives greater credibility to the presentation than simply talking about water quality, and conducting an experiment in front of the audience is a powerful presentation tool.

How do you want to be perceived as a presenter? Think about presenters you have seen that you considered highly skilled and effective. What aspects of the performance did you most admire? There may be techniques or formats that you will want to develop within the context of your own presentation style, although caution must used to ensure that the presentation remains authentic and true to you as the presenter. For example, if you desire to be viewed as a highly credible expert among a group of staid academics, but your personal presentation style falls to the 'prop comic' end of the spectrum, you may want to temper the use of squirt guns and canned snakes (although one or two humorous touches will make your presentation your own).

In the past, most texts on presentation styles would have suggested that certain styles are required for certain situations. That is, a presentation to a professional group would require a focus on facts and statistics with less emphasis on humour or personal reflections. Today, this may or may not be true: there is much more flexibility in presentation styles and the audience may be receptive to something new. However, the presenter's style must respond to:

- *The client's request.* What are they expecting of the presentation? A personal testimonial? Cold hard facts? What do they think you will be doing?
- *The number and composition of the audience.* Is it a large group, where your presentation will be broadcast on a giant screen? Are you speaking to a few professionals in a small venue? Generally, the larger presentation would require visual aids and likely more generalization to the presentation (see Chapter 15).
- *The time frame.* How much time do you have? Does the client expect a three-hour lecture or a 10-minute overview?
- *Your desired impact.* How do you want to be viewed? If the topic is very serious, you may want to assume a presentation style that speaks to facts and documented evidence over one that will keep them laughing.

Ultimately, the goal of the presentation *is to be understood*. The development of your own professional presentation style is a way of ensuring that your message reaches your audience.

Practical tip. Do not think, 'How am I going to survive this?' Instead, ask yourself, 'How am I going to be brilliant today?' and let your presentation flow.

Chapter Review

This chapter considered the kinds of questions a presenter is most likely to receive from the audience. In addition, techniques for responding to different question formats were provided. As the presenter, you have control over the presentation. Retain this control to ensure an effective presentation. We have also seen in this chapter that there is no single right way to present, but that techniques and methods can be used by anyone to be a more effective public speaker.

Review Questions and Activities

1. What type of question is most commonly asked? As a presenter, how can you respond to this form of question?

2. How should a presenter respond to an audience member asking a question that is clearly off-topic?

3. What is the role of the audience?

4. What is the ultimate goal of any presentation?

5. How do you want to be viewed as a presenter? Take a few minutes to consider your presentation style and see if it accurately represents how you want to be perceived by the audience.

PART IV

Illustrations and Mapping

The basics in document layout and design are considered in this section. These graphic guidelines are discussed in the context of figures, tables, static presentations, and PowerPoint presentations. We also consider when you can break the rules in regard to visual presentations.

Part IV examines, as well, the value and use of maps in presenting information. The power of the map to communicate data is considered, as are the basic elements of mapping.

Chapter Fourteen

Visual Design

Good design helps any document or presentation. The judicious inclusion of graphics and illustrations, the way text is arranged on a page, and the use of colour can do more than make a document or presentation appealing—they can affect the perception the audience has of the author or presenter. While a page of single-spaced, justified black text (Times New Roman, font size 12) can give information, a well-designed presentation can add to the reader's understanding and information retention.

Given the range of software and presentation tools available to presenters, audiences expect a high and increasing level of visual design in any professional work. As an audience, we expect to be visually entertained as well as informed. New standards are set daily; for example, not so long ago, it would have been highly unusual to insert a video into a PowerPoint presentation, but with simplified technology, videos are now frequently used to emphasize a point or showcase a project. Professional speakers are now using wall-sized multiple screens for presentations that are highly kinetic, with shifting visuals and perfectly timed effects.

Along with being up to date on presentation technology and software, we expect that the presenter will know the basic principles of visual design. This chapter outlines five design principles that can make any presentation or document more visually appealing for and have a greater impact on an audience.

Why Is Design Important?

Because of the ease of typing and editing text, most documents and presentations are developed first as word-processed documents. On occasion, the writer will choose a different font from the standard Times New Roman 12-point that better represents the writer or the topic (for example, often **Comic** is used as a font for text directed towards elementary school children). Less frequently, the writer will consider the visual impact of the text and apply a few principles of visual design to produce a document that is both informative and visually arresting.

Box 14.1 Two Rules for Graphics and Illustrations

Rule #1: Just because you can does not mean you should.
    If a graphic or illustration does not add to the meaning of the text or slide, do not use it. Extraneous, unrelated graphics are distracting and can be annoying to the reader.

Rule #2: Universally accessible clip art should be used sparingly, and only if it complies with Rule #1.
    If the graphics used are available openly on-line, this means that anyone can use them. The same image in your newsletter may be used by any number of individuals and organizations, potentially for uses that are at cross-purposes to your application.
    If the artistic abilities or funds are available, consider creating your own graphics and illustrations and copyrighting them for your own use. The development of a visually appealing logo and themed graphics that create a 'brand' for a company is often one of the first expenditures for new businesses.

In many universities, students are required to produce term papers and essays of prescribed length, spacing, margins, etc. The visual appeal of the document is given limited consideration; instead, the text itself is to carry all meaning. While this approach to presentation is useful in a university setting where the text and content are the intended focus, the same is not true in most business settings. In the outside world, the text, content, *and* the visual appeal of the document are all important. In fact, depending on the purpose of the communication (e.g., is it an internal memo or an advertisement in a major magazine?), the visual appeal of the document may be paramount.

While some businesses have graphic design professionals on staff or have graphic artists and web designers work for them on an occasional, contract basis, it is also highly possible that you may be it: you may be designing your own website, newsletters, letterhead, reports, and presentations without any professional assistance. Adhering to a few simple guidelines can make every page or presentation more effective, more striking, and more easily understood by the intended audience.

Perceptions of a Page

First, it is important to consider how we have been trained to view written documents and presentations (from a Western perspective). Consider how you read a page of text: your eye travels first across the top of the page, reading the heading if there is one. Next, your eye likely flits to the next most prominent feature on the page, such as a diagram or photo. Only after you have quickly assessed what is on the page besides straight text will your eyes return to the top of the page to read the text.

When first viewing a document that contains more than a straight page of text (including anything that breaks a solid pattern of words on a page, such as graphics, a photo, or even changes in typeface), research in eye movement has shown that Western readers tend to follow a 'Z' pattern across a page: the eye moves across the top of the page first, then travels in a diagonal to encompass the entire page, then across the bottom of the page.[1] This eye movement varies if there are features on the page that draw our attention. We are attracted to high contrast elements and things that are different (larger in size or font, brighter in colour, or of high graphic weight, such as a photograph). We expect the most important information, as shown in Box 14.2, to be in Quadrant 1, the top left corner. Quadrant 4 is an anchor space, the space where our attention last focuses on the page. Quadrants 2 and 3 tend to be less viewed when tracing eye movement, unless elements in these areas attract the viewer's attention. We tend to follow items that are along a straight line, and will skim among items that are the same, mentally grouping these items.

Knowing this, we can make the arrangement of features on a page more thoughtful and planned. The most important information should be contained in Quadrant 1, potentially emphasized by using visual design guidelines to add to the

---

Box 14.2 Eye Movement

Quadrant 1	**Quadrant 2**
Quadrant 3	**Quadrant 4**

visibility of the items. Secondary information should be contained in Quadrant 4 and should support Quadrant 1. Quadrants 2 and 3 are the location for background, filler, and less critically important information, such as a photo for colour or a diagram that adds to the text but is not imperative to its understanding. Thus, if we know how most people view a document, it makes sense to arrange the elements on the page to suit these cognitive patterns. It also makes sense to employ methods (like the use of contrast and connectivity) that create a visually appealing display and add to viewer understanding.[2]

It is important to know that not every reader will view every word on a page of text. Many viewers skim over the information, reading only what catches the eye and forming a perception on their understanding of the page. Any document or presentation will be improved by considering five easy-to-apply guidelines: *contrast, consistency, connections, closeness,* and *colour and symbols.* These guidelines will help to ensure that the viewer understands and retains the most important elements of your presentation.

Contrast

Highway signs are black text on a yellow background. Why? Because the two colours are far apart on the colour spectrum and show a high level of contrast. Black text on a white page also shows a high level of contrast, where cream-coloured text on a white page does not.

Contrast helps the viewer to understand information quickly. We expect things that are important to be visually larger, darker, and more intense. If there is text in a large font and a small font on the same page, we will read the large text first. We notice bright colours, large shapes, and things that are different.

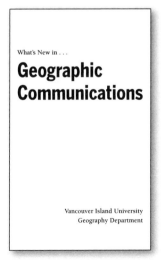

Figure 14.1 Using Boldface to Attract Attention

Ways to Use Contrast

- Use different fonts for different information (one for titles and one for text), but limit the use of different fonts to two per page. One font could be a serif font (with tails on the ends of letters, like **Times New Roman**), while the other is sans serif (that is, without serif, like **Tahoma**). As a general rule, a serif typeface is best for text because it is easier to read.
- Make the two font sizes different. If they are both 12-point, for example, there will not be enough contrast. However, don't go overboard in this regard—then you are basically telling the reader/viewer that the smaller type isn't worth reading.
- Use boldface and/or italic to highlight important text, but again, there can be too much of a good thing. Too much bold gives the readers the sense that you are trying to tell them how to think and react, and too much italic is difficult to read. You wouldn't want to give all or much of an oral presentation in your loudest voice with flamboyant gestures; neither should you on the printed page. Less can be more.
- If colour is used, keep contrast high. Make important text, such as headings, stand out from the background colour and the regular text of the document. Be brave but not garish in colour choices—it can be visually exhausting to view slide after slide of black text on bright yellow slides, for example.
- Stand back and view the page or slide. What draws your eye first? Second? Use contrast to pull the audience to the most important information.

Consistency

Consistency is creating a theme or scheme for your work. Creating a pattern that the audience can follow helps to hold less significant elements in the background while still giving them meaning (imagine a slide presentation where every slide was a different template, colour, and font combination—it is unlikely this would be successful, unless your product *is* slide templates and font packages). For example, a slide template that contains a small image of the company logo in Quadrant 3 is a good means of keeping the image in the viewscape of the audience while allowing the audience to focus on the important elements of the presentation. Other elements can be held consistent, like a font that is used in all presentation materials (including letterhead, business cards, and the font on a slide).

   The consistent use of colour choices can also speak volumes about the presenter: although caution should be used in reading too much meaning into colour as individuals have different perceptions, certain colours like green are associated with nature and the environment, while blues may be viewed as more corporate. A consistent use of colour either as background or in the elements of the presentation will impart a message to the audience. Bear in mind, however, that approximately a sixth of the population has some form of colour-blindness, and that colours do not suggest the same meanings for all people or all cultures.

Whether people notice colours and fonts consciously or subconsciously, they will group this consistent information into themes of meaning and a total perception of the information.

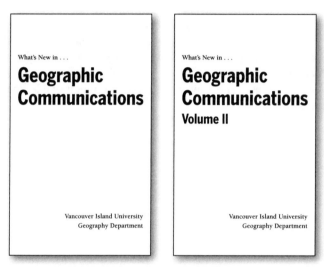

Figure 14.2 Maintaining Consistency

Ways to Use Consistency

- Create a hierarchy in font size and typeface for titles (main title the largest, then second level for the next, and smallest for text).
- Keep elements the same that should be the same. For example, keep all sidebars of information in the same font and the same approximate location on the pages where they are used.
- Create a pattern of reoccurring elements: page numbers are always in the same place, the company logo is on each page, and fonts are the same among different documents. The development of a look or theme assists the audience in recognizing 'what's next' and allows them to focus on what's important.
- Create obvious patterns but allow elements occasionally to break the pattern to create interest and keep the audience awake. For example, if you have presented slide after slide with the same overall format (title at the top and a table, photo, or graphic below), shake up the presentation with a new element that is relevant but contrasts with the expected pattern.

Connections

Eye movement studies show that the eye traces elements that line up or create a hard edge on a page. That is, we seek out alignment and visually favour items that appear to be placed with a connection to other items. These connections create flow and coherence on the page, and allow the viewer to trace the edges of the information and then refocus on the elements that are more important. This design guideline is perhaps easiest of all to achieve, and requires only the consideration of two issues: where to justify text, and what should be grouped together.

Word processing offers four alignments: left, right, centre, and justified. We expect most text on a page to be either left aligned or justified. Titles can be left, centre, or right justified; centre and left are most common, while right justification can appear more interesting.

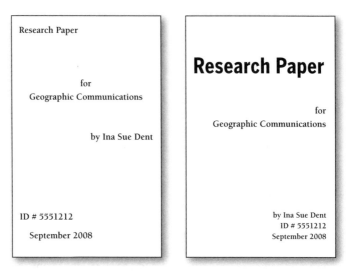

Figure 14.3 Improvements with Connections and Closeness

Ways to Use Connections

- Items can be grouped vertically and/or horizontally on a page, and every item on a page should line up with another element.
- Create edges on the page—not with vertical or horizontal lines, but allow the eye to trace the edge. This is more subtle and professional than drawing a frame around the text.
- Use connections as a technique for keeping like items together, along with your consideration of contrast and consistency.
- Keep all headers of the same value at the same location on different pages, maintaining left, centre, or right justification throughout the document or slide presentation.

Closeness

Strongly related to connections are the guidelines around closeness. Simply, keep elements that belong together in close proximity, and keep unrelated elements apart (see Figure 14.3). Use the white space on the page or slide to separate elements: in a document, leave one or two extra line spaces between sections to separate them visually. Use page breaks to create an even clearer separation. On a slide, group background elements together (like the company logo and the slide number) in an inconspicuous place, and keep the titles and text higher on the slide. Negative space, i.e., the empty space around the type and visuals (the positive space), is important. The negative space allows the viewer to keep elements in logical groups, which facilitates greater and faster understanding of the information presented.

Colour and Symbols

Most documents and presentations benefit greatly from the use of colour. Colours evoke meaning—most North Americans would equate green with the environment, the outdoors, and lowered stress, while red is equated with excitement, vibrancy, and growth. The use of colours, then, should be carefully considered. Be certain that the message and meaning represented to the viewer are consistent with your goals and image. For example, the use of extremely bright primary colours (along with an unusual font) in a financial products newsletter may not engender investor confidence. Nor would a lot of bright reds and oranges (hot colours) be suitable for the newsletter of an environmental group, unless perhaps the newsletter is warning of ecological Armageddon.

Symbols also impart meaning. A symbol can be a picture (like a happy face) meant to cause a reaction or emotion in the viewer (a smile or happiness) or it may be a representation of the true appearance of a feature (such as the use of a wavy line to represent a river on a map).

Symbols carry the danger of misinterpretation. A poppy is generally associated with Remembrance Day in Canada, but may have an entirely different meaning to a drug enforcement officer in an opium-producing nation. Use caution if the document or presentation will be viewed by a range of audiences or if the audience composition is not known, and select symbols that will have a neutral value for most viewers (like dots to show location and unremarkable lines to show linear data).

Ways to Use Colours and Symbols

- First, use colour when possible. The use of colour will make your document and presentation more memorable.
- Colour and impressions are closely related. The colours you choose will offer an impression of you, the writer or presenter.
- Garish colour combinations or the use of too many colours appears unprofessional. When unsure of what colours to use, one rule of thumb suggests

staying with the colours that are representative of the natural environment at your site.

- Consider traditions in the use of colour, particularly for mapping. On many zoning and development maps, yellow is used for residential, red for commercial, green for park space, and blue or purple for industrial lands. While these are not axioms in mapping, they may provide a starting point for colour selection.

- Once a main colour is selected, develop a scheme that suits the document or slide. Maintain this scheme throughout (and potentially over many documents or slides if a corporate look is adopted for communications).

- Consider the use of black and white—black adds crispness to any page and remains highly legible for blocks of text on a light background. White adds space and openness to a page or slide. Both should be used to give depth and focus to a colour-filled presentation. White type on a dark background, however, is difficult to read and should be used sparingly, if at all.

- For symbols, be certain the symbol used will not be negatively interpreted or associated.

- If colour and symbols are combined to show a range of occurrences of a phenomenon, it is usual to make the darkest shade or largest symbol representative of the highest number of occurrences or density, and the lightest or smallest to show the fewest or least dense. Any progression should follow a rational order, and break points should be carefully determined to fit the data set.

- It is also possible that colours and symbols are used to show completely different data sets. In this instance, it is important that the colours be diverse enough to illustrate the different data classes.

Bringing It All Together

Examine the sample pages in Figure 14.4. Which page is easier to understand, more appealing, and more likely to have an impact on the intended audience?

You may say the page on the right is more informative—it certainly contains more information—but it will not be informative if it is never read.

The placement of text and graphics on a page is critically important to ensuring that the intended message is understood by the reader. Again, the point of writing and presenting is to be understood: a good presentation makes it easy for the reader to understand the purpose of the text, and an appealing visual presentation increases comprehension and the likelihood that the text will be read.

Whenever possible, build the document or presentation for the intended audience. Simple changes, such as increasing the text size for a presentation to a seniors' group or decreasing the amount of text for a presentation to elementary schoolchildren, can help to ensure that the presentation is understood. Changes in colour or graphics used, as well, will add credibility for the audience. A switch in the background colour of slides in a presentation from corporate blue to environmental green would be recommended if the same presentation is being given to a right-wing business group one day and an environmental activist group the next.

India

Population: over 1,000,000,000 in 2000
Population growth rate: 1.71%
Birth rate: 25.91 births/1,000 population
Death rate: 8.69 deaths/1,000 population
Sex ratio at birth: 1.05 male(s)/female
Infant mortality rate: 63.14 deaths/1,000
live births
Life expectancy at birth total population:
62.9 years
Languages: English, lingua franca; Hindi,
national language and primary tongue
of 30% of the people; Bengali (official);
Telugu (official); Marathi (official);
Tamil (official); Urdu (official); Gujarati
(official); Malayalam (official); Kannada
(official); Oriya (official); Punjabi (official);
Assamese (official); Kashmiri (official);
Sindhi (official); Sanskrit (official);
Hindustani, a popular variant of Hindu/
Urdu, is spoken widely throughout
northern India.

- Will surpass China as the world's most populous nation by 2050
- English is the language of government and business
- Low population growth rate but a high birth rate
- Life expectancy rising

☼ **World Consulting Ltd.**

Figure 14.4 The Presentation of Information
Photo: Rachel Bartles

When Not to Use the Guidelines

The guidelines outlined in this chapter are intended to assist in the development of a presentation or document—once templates are established and a look is created that best represents you or your organization, filling in the blanks becomes easier. Also, once these guidelines are internalized and used in the development of documents, time can be saved in production and comfort can be found in knowing that information is being presented in a way that is more likely to be understood and retained by the audience.

However, a guideline is a recommendation, not a law, and there are times when guidelines should be tweaked or ignored completely. There are no guidelines for breaking guidelines; the development of new visual impacts comes from an understanding of the principles of visual design and the creative breakdown of the expected.

Once you have prepared a slide presentation or document, consider if there are places where something out of the ordinary would add interest, life, and unique character, bringing in the unexpected. Your goal remains the same as the goal of the writer discussed in Part I and the presenter considered in Part III: to be understood. Stepping outside the guidelines to ensure greater understanding is doing a favour for your audience and will allow the document or presentation to meet your new professional standards.

Chapter Review

This chapter has considered elements of visual design that can be applied to any document or presentation. The Five Cs were reviewed: contrast, consistency, connections, closeness, and colour and symbols. When not to use the rules was also considered—understand the rules to break them intentionally.

Review Questions and Activities

1.   What is the difference between contrast and consistency?

2.   What issues might impact the choice of colours or the use of colours at all?

3.   How do most Western readers view a page of text or a slide? What is the eye pattern, and how can this information be used to the author's advantage?

4.   Select any document you have produced in the last six months. Revisit the document, using the guidelines discussed in this chapter. What improvements would you make?

5.   Why is design important? What other rules of good design do you incorporate into your work?

Chapter Fifteen

Visual Aids

An effective presentation style combined with good use of visual aids can elevate a presentation and make it more memorable for the audience. Visual aids should be considered a tool, just like a hammer or hacksaw. They assist in reaching a result. A PowerPoint slide that shows a map of a study area offers a short cut to audience understanding—while the study area could be explained in words, the image is likely more powerful and better guarantees a similar level of understanding across the audience. Visual aids should only be used when they will be effective: if they will help to reinforce a point or aid understanding, then there is a place for the use of visual aids in the presentation.

Box 15.1 The Use of Pointers

If you choose to use a technical aid such as a laser pointer or retractable pointer (the ones that collapse to the size of a pen), the cardinal rule of presentations applies: *practise, practise, practise.* An audience will quickly lose patience with a presenter who cannot manage his or her presentation tools.

When laser pointers were first widely available, many presenters (to ensure that they appeared leading edge) leaped on the bandwagon and audiences everywhere suffered motion sickness from viewing a violently bouncing red dot, pointing at everything and nothing.

The overuse of laser pointers has made them much less common in business presentations. If you are projecting a complex object like a world map, a laser pointer may be useful for pointing out specific features. Keep one available in case it is needed, but use it only in very limited circumstances. You do *not* need to follow text along with a laser pointer.

Stick pointers (either a long wooden pointer or the retractable pointers) are also seldom seen: they can be useful in limited circumstances, such as pointing to objects on a blackboard in a school classroom, but are too limited in use for most business presentations.

Drawing a diagram on a chalkboard or flip chart or displaying an overhead or slide is a way of bringing a second layer of information to a presentation: the presentation goes from being largely auditory (where learning is based on listening to the presenter) to both auditory and visual (where the audience listens to the presenter and gains understanding from the visual aids). Visual aids can add interest to a presentation and increase the audience's ability to remain focused on the presentation and learn from it. The use of visual aids takes advantage of many people's primary method of learning—through observation.

While the percentages differ slightly in different studies, research in memory retention has determined that after 24 hours audiences retain:

- 10 per cent of an auditory message;
- 35 per cent of a visual message;
- 55 per cent of an audio *and* visual message.[1]

The use of visual aids can be effective in reaching an audience and having them remember what you tell them.

However, just because you have the technical ability to use or access a visual aid does not mean that you have to or should use it. In fact, a well-presented, enthusiastic, knowledgeable presentation using a discussion or lecture format can be infinitely more informative than a slick multi-media show. Visual aids are tools that help you get a message to an audience, whether that audience is seated in front of you, reading your journal article on demographic and occupational change in Sudbury, or finding your flyer or brochure with the rest of the morning mail. Use these tools when they will help you to build a better presentation and help your intended audience to understand and remember what you have to say, but not simply to try to hide the fact that your presentation lacks substance. In this chapter we examine the kinds of visual aids as they apply to public in-person presentations.

Kinds of Visual Aids

The most commonly used visual aids are overhead transparencies, flip charts, PowerPoint slides, and videotapes and DVDs (film slides are increasingly rare). Which one is most effective? The following outlines the benefits and limitations of each aid.

Overhead Transparencies

Overhead projectors are seldom used in business presentations, but they remain in use in some settings, such as school classrooms and universities, and in some places in the world where computers and projectors are not widely available. The overhead has the advantage of being able to be seen by a relatively large number of audience members, and the technology is fairly inexpensive and highly reliable. They are also useful as a backup to a PowerPoint presentation if you are not certain that the venue will have a projector available or if that projector will be compatible with your laptop.

Transparencies are inexpensive to produce, and can present text, charts, or drawings effectively. A further advantage is that the venue may remain well-lit; an overhead projector generally does not require that the room be dimmed to view the transparency. The presenter can remain connected to the audience through visible eye contact and non-verbal cues that keep both the presenter and the audience in the presentation. It is also possible for the presenter to write on the transparencies, diagramming an issue and noting ideas. This requires a certain skill level: hopefully, the presenter has artistic or penmanship abilities that allow for clear understanding from audience members.

Overhead transparencies do have some limitations:

- The image projected may be too small or light to be seen from all points in the venue.
- The presenter is trapped next to the projector—it is difficult to move any distance and still be able to change transparencies.
- The image is often distorted if the head of the projector is not completely square to the screen.
- Only a limited amount of text should be contained on the transparency (as is true for any slide).
- Photographs generally do not present well on an overhead projector.

Box 15.2 Flip Chart Tips

- Practise writing clearly and legibly; there is little point in recording information that the participants cannot read and that you will be unable to understand when later transcribing the flip chart.
- Size your writing so the participants can read the material—check from the back of the presentation space.
- Use colour whenever possible. Keep the main text in black, but use colours and highlighters to group information and add visual interest. (Remember, though, not to rely solely on colour, since a surprising number of people have some form of colour-blindness.)
- Know in advance if you can tape flip-chart pages up in a room—not every venue will allow you to put tape on or pins in the walls, and some wall finishes (flocked wallpaper or fabric) will not allow tape to stick.
- Know your materials—do not use felt pens that will soak through the paper, or tape that has limited stickiness. Watch for pens that are very smelly: increasingly, people have environmental sensitivities to smells, and felt pens can be toxic to persons with strong allergic reactions.
- If possible, purchase flip-chart paper that has pre-stick strips along the top edge. They appear neater when attached to a wall surface than masking tape.

When should you use overhead transparencies? If you have a few simple diagrams and the projector is available, overheads can be a simple and effective means of adding to your presentation. Your presentation may also appear refreshingly 'old school' to the audience, especially if you are presenting along with others who allow the technology of videotapes, DVDs, and PowerPoint slides to overtake the message in their presentations.

Flip Charts

As with overhead transparencies, flip charts are an inexpensive means of presenting information to a group or recording information for a group. They are most often used when a large group breaks down into smaller groups, with each group having the same conversation or responding to the same questions. Generally, one person records the group's comments on a flip chart and reports back to the larger group, or one person can be assigned as the recorder and another as the presenter.

Flip-chart papers can also be prepared in advance. If you are presenting with a flip chart, graphs, diagrams, and main points can be drawn or plotted on a large-format printer, then arranged on the flip chart before the presentation (if this presentation technique is used, place a blank page between each presentation page to ensure that the later page is not visible through the previous page). The presenter then flips two pages at once (you can even tape the corners of the pages together to ensure the pages flip smoothly). The papers can be reused for other presentations; if you are likely to present the same information again, have the papers laminated with a matte finish (to avoid glaring reflections).

Flip charts have some limitations:

- Their use should be limited to small groups—with more than 10 it becomes difficult for participants to sit close enough to be able to review the information.
- A level of artistic ability and good penmanship is required—messy or illegible flip charts do not add to the presentation.
- Flip charts can appear awkward and unwieldy, and the presenter has to be conscious of avoiding the legs of the stand and competently flipping the pages over the top of the chart (no small task for a short presenter).
- Depending on left- or right-handedness, the recorder may be standing directly in front of the flip chart (with her or his back to the group) while writing, making it difficult for presenters to see the information.

Also, a caution to the recorder: you are recording the group's ideas and comments, and not necessarily your own. If you are a participant and also the recorder, be sure to vet your ideas to the group before placing them on paper, and be sure you are capturing the ideas of the participants adequately as you write them down. The group will quickly lose interest if they observe that their comments are not being recorded or are overtly editorialized to fit the recorder's agenda.

Flip charts are also increasingly rare in business presentations. They lack the professionalism of other visual aids, and someone is always left with the difficult task of transcribing pages of comments. More often, if information needs to be recorded, a better method is to have a computer/projector set up and a person assigned to record and group information as it is obtained from the participants. The information can be projected on a screen, or quickly printed and given back to participants, ending the need for transcribing text from the flip-chart pages.

PowerPoint Slides

As a presenter, there are many ways to misuse a PowerPoint presentation:

- Unrelated information on a single slide.
- Too much writing on each slide.
- Using the slides as an organizational tool for the presenter instead of considering their effectiveness for the audience.
- Moving too quickly through the slides, limiting the audience's ability to understand their contents.
- Overusing show techniques (such as moving text or sound effects). One or two may be effective (in certain appropriate venues) while more are distracting.
- Overusing scanned or on-line images that have little relationship to the topic (the 'just because you can' error).
- Showing bullet point after bullet point, with little point to the presentation.
- Letting the 'slide show' distract from the flow of the larger presentation (refusing to answer questions or interact with the audience because it will impact the pre-set timing on the slides).
- Believing that slickness replaces knowledge.

A cardinal rule for presentations that use text on a PowerPoint slide: never read to the audience. In most situations the audience can read, and they don't need you to read to them. If you are making a presentation where you are not certain about the reading level of the audience (e.g., a presentation to children, in a situation where literacy rates may not be high, or in a context where English may not be the first language of many in the audience), why would you be presenting words on a screen?

There is only one instance when reading from a slide can be used as an effective presentation technique: when reading is critical to reinforce a key point. Reading aloud a quotation as it is presented to the audience on screen can bring nuances of meaning to the quote that may have been missed by the audience. Reading provides emphasis and focus; used sparingly, it provides a 'change in voice' for the presenter and brings a different energy to the presentation.

An effective presentation focuses on slides that add meaning to the presentation. Simple graphics and charts help the audience to understand numbers and relationships. Photos can be very effective when used to support the words of the presenter. The use of pictures of people, places, and things that are known to

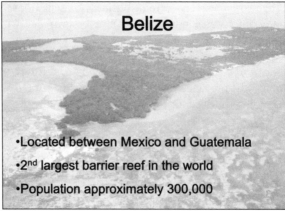

Good

Facts on Belize

•Belize is located between Mexico and
Guatemala, on the Caribbean side of
Central America and is at latitude and
longitude of 17 ° 15' North and 88 ° 45' West
•There are approximately 250,000 to
300,000 people in Belize
•About 40% of the nation is covered as
protected areas, which means it has some
level of protection against development
•Belize contains the second largest
Barrier Reef in the World - only the Great
Barrier Reef in Australia is larger

Bad

Figure 15.1 Good and Bad PowerPoint Slides

the audience can be powerful and add layers of meaning to a presentation. For example, iconic images like the Eiffel Tower or the Statue of Liberty are recognized worldwide. However, there are two dangers with the use of familiar subjects:

1. Not all members of the audience may give the same meaning to an object (an image of a crowded urban street may appear exciting to some and fearful to others; people in uniform often have different and opposite meanings to different people, depending on ideological perspective and personal experience).
2. It is risky to screen locally known subjects in a negative light. An audience may take offence at the image's inclusion (particularly if the presenter is not from the community) and the presenter will, at best, have a distracted audience; at worst, the audience will have been irrevocably lost.

Box 15.3 PowerPoint Tips

1. The rule of three: if you are using text on the slide, do not have more than three bullet points. The slide is meant to present an outline for the audience, not every word you are going to say.
2. Better yet, do not use bullet points at all. Use images that tell a story.
3. Do not use 'pixelly' photos or poor-quality images. If using digital photos, be certain that they retain their quality when resized. Also, do not 'borrow' images from websites without considering copyright laws. Just because it is easy to 'borrow' does not make it right. Use non-copyrighted materials or your own original photos.
4. Sans serif fonts (that is, fonts that do not have 'tails' at the ends of letters) are more easily read on a screen than serif fonts.
5. Create your own templates. Do not use the same templates that everyone else uses. Keep contrast high to aid in legibility.
6. You do not need to remain glued to the computer when using PowerPoint. Purchase a wireless slide advance tool (sometimes a wireless mouse will work) to allow movement around the presentation area. If you are using a microphone, confirm in advance that it also is wireless so you can move and gesture easily.
7. Resist the urge to turn off all the lights in the room. Be sure that the photos and graphics used are of sufficient quality that the room can remain lit (and the audience awake).

Consider if there are opportunities for the audience to misconstrue or take offence to any aspect of the presentation—a good technique to implement for any presentation is to test it with the event organizers or with people who will be familiar with the audience. Time spent on understanding the audience will always lead to a better presentation.

Video clips may also be inserted into a PowerPoint slide show. This can be very effective as it introduces a new medium to the presentation. However, care must be taken to ensure the clip is relevant, of good visual quality, and not too long. A blurry video filmed from a hand-held camera might make millions at the box office, but it will not be effective in a presentation. Audience expectations are different, and the contexts are likely quite different.

A final option is an audio slide show, where images are presented in an array set to music. If you use this technique: be certain that the timing works. It is extremely difficult to time a presentation to a pre-set show, and doing so removes any opportunities for a spontaneous comment or interaction with the audience. The presentation changes from an exchange to a show, which can be useful if you are playing a video loop at the side table at a conference, but it can feel contrived as a presentation. A second caution is to not overplay the transitions between slides. An effect that is interesting the first time is not so exciting 100 slides into the loop.

Videotapes, DVDs, and Web Feeds

Professionally produced videotapes and DVDs are widely used and highly effective visual aids. Their advantages are obvious—high production values and professional actors. Videotapes and DVDs can be used to program an entire presentation or to support a presenter's remarks with carefully selected clips (if copyright is obtained). Their disadvantages are also obvious—extremely high costs and long time frames are two factors that make the production of videotapes and DVDs beyond the realm of the possible for most presenters.

The ubiquity of video equipment combined with editing software may make a self-produced tape of interest to the presenter. The newest forms of presentations incorporate RSS feeds off the Internet—these feeds can be updated almost instantaneously and can show live events from anywhere around the world. Combination formats are also possible, with web feeds embedded into a PowerPoint presentation making the presentation 'up to the minute' in content.

Practical tip. A danger lurks in letting the technology take on greater importance than the message. Unless the product is very good, it is unlikely to add to the audience's understanding of the issue being discussed. They are more likely to be distracted by aspects in the production that are less than expected (or worse, by unintended gaffes). Unless time and budget are of little importance, a self-produced product is not recommended.

A Final Word on Visual Aids

If you choose to use a visual aid, it is absolutely necessary that you know how to use it and how to troubleshoot should problems arise:

- Arrive at a venue early, leaving enough time to implement a 'Plan B' if the needed equipment is not working or unavailable.
- Check the room set up to make sure that it is as expected for your presentation. Know where light switches are located and how to turn on the mike.
- Practise your presentation with the equipment you will be using once the audience arrives.

Use visual aids as a tool to make your presentation as powerful as possible.

Chapter Review

This chapter has considered the use of visual aids as they relate to public presentations. From overhead projectors to RSS feeds, technology can supplement what you are presenting to provide greater interest to a presentation. The emphasis, however, is on *supplement*. Do not allow the technology to overwhelm the presentation. If you find you are spending more time on the format of the presentation than on the content, you have likely let the format take over your intended message.

Review Questions and Activities

1. Compare and contrast the visual aids discussed in this chapter, stating the strengths and weaknesses of each format.

2. Think of presentations you have seen in the last year. What visual aids were used? Were they effective? What changes would you recommend to these presenters?

3. What is the cardinal rule for a PowerPoint slide?

4. How does the 'Plan B' rule apply to visual aids?

5. For your next presentation, use a visual aid format that you have not used before. If you rely on PowerPoint, try to use a flip chart instead. Learn to develop a series of presentation methods that you can rely on to fit different venues and circumstances.

Chapter Sixteen

Presenting Numbers

Numbers are a shortcut to a clearer understanding, are a way of illustrating complex relationships, and can be used to assist in the audience's understanding of a situation. For example, news reports on earthquakes will immediately note the Richter scale measure, as it is a generally understood concept and gives the public some understanding of the extent of the event. Even if the audience is not comprised of geologists, most people understand that an earthquake of 7.6 will potentially have more impact than an earthquake of 3.2, depending on the location and ground conditions.

However, most people can retain or interpret no more than a few numbers presented at one time, whether in a single sentence or on a slide. As a presenter, once you understand this axiom, presenting numbers becomes simple—the presentation focuses on the audience and their understanding of the numbers, and *not* on the numbers themselves.

As the presenter, your task is to make the information understandable, not to baffle the audience with so much statistical detail that your point is lost, either because your presentation is incomprehensible or, worse, boring. When presenting numbers, then, three rules should be considered.

Rule 1: Know your audience. Who will be the audience for the presentation or who will be reading the text? A university statistics class will have a greater ability to understand complex charts and figures (and may even have a small amount of interest in an in-depth analysis) than most of the general population.

Rule 2: Make comparisons relevant. Again, the audience is the guide. Is the comparison relevant to the reader/viewer? Do they 'get' the comparison? Comparisons should be used to create an accurate mental image that will be useful in your understanding of the discussion that follows. For example, consider the following imaginary statistical comparisons:

- First-year university students in Canada consume enough coffee to fill 10 Olympic-sized swimming pools daily.
- The size of the newly created Baltic state will be twice the size of Arizona.

- The amount of money annually available in scholarships in Canada is roughly the same as dollars spent on imported beer in student union pubs across Canada.

The relevance of the comparison needs to be carefully considered: most readers could conceptualize the size of a swimming pool, but will all readers be familiar with the size of Arizona? Would it not be more relevant to list the dollars available as scholarships by province or area of study rather than comparing the national total to a completely unrelated statistic on imported beer? This, of course, might depend on whether your subject is scholarship funding or imported beer, but in either case a number would help. The use of comparisons must be carefully considered against the tests of relevance to the audience and whether or not the comparison is more spurious than serious.

Rule 3: Know when numbers are worth 1,000 words. If a table or graphic is used to explain a key concept, it is not always necessary to re-explain the concept in text. That is, if the graphic is effective, the text becomes redundant. Carefully consider if more needs to be said about the numbers presented or if they truly speak for themselves.

Issues to Consider in Using Numbers

When presenting numbers, there are four issues to consider: accuracy, understanding, detail, and necessity.

Accuracy of the Source

Are you certain that the numbers you are presenting are correct? Even sources considered reliable, such as government statistical organizations (like Statistics Canada or the US Census Bureau), recognize that their data will never be 100 per cent accurate. Under-representation of certain populations (such as the homeless) impacts census data, and misreporting by respondents (either inadvertent or intentional) also affects the results. Sampling sizes and the sampling methods also may lead to questions on the veracity of data. As a rule, never present information unless you are certain of the source and can defend the source as reputable. Agenda-based websites or 'think-tanks' that produce statistics from unverifiable survey methods or sources may be seen as questionable or even fraudulent by the audience.

What can be considered trustworthy information? In general, government-collected census data in Canada and most other Western democracies can be considered reliable because the collecting agency has no particular stake in presenting a predefined viewpoint, and an attempt is made to use the best available survey methods and samples that reflect the entire population under study. From an academic perspective, credible studies are usually those that have been vetted through peer-reviewed journals or replicated by other researchers.

Credibility becomes 'fuzzier' when the information is collected by a stakeholder or interest group that may have directed the survey results, either through the survey questions, the survey methodology, or though the analysis of the results. As discussed later in this chapter, the statistician has the same abilities as the map maker to alter the viewer's perceptions in a number of ways:

- Questions may be phrased to lead a response by only offering choices that suit the surveyor's predetermined result.
- Samples may be selected in a non-scientific manner (for example, by only polling subscribers to a right-wing magazine).
- Data that do not meet the predetermined result may be classified as an error (examples have been seen in democratic nations, including Canada, where election workers were more likely to mark ballots as 'spoiled' when the vote was for a candidate not supported by the election worker).
- Only data that meet the predetermined result are presented.

If the source could be considered questionable by the audience, your use of the information (along with the data itself) may be seen as suspect. Unless you are prepared to defend the organization along with the data, you may want to choose a more neutral and reliable source. Also, be certain that you are not propagating false statistics: the credibility of your presentation or document (and even your personal credibility) depends on the veracity of the numbers presented.

Understanding the Numbers

A second issue is ensuring that you understand the numbers. If you are presenting statistics, graphics, or charts obtained from a reliable source as part of your presentation, be certain you fully understand both the numbers shown and the meaning behind the numbers.

For example, the table in Box 16.1 shows a ratio of richest to poorest populations in a series of nations. If you are using this information as part of a presentation, could you fully explain the table? Invariably, if you cannot, someone in the audience will ask a question on it and you will lose credibility as a presenter. Why would the audience trust a presentation from someone who cannot explain information they have chosen to present?

Level of Detail

If too many numbers are presented to an unnecessary level of accuracy, the point of the presentation becomes lost in the numbers. On occasion, complete accuracy to many decimal places is required to ensure that the audience has the necessary information to understand your presentation. If you are presenting at NASA on the re-entry of the space shuttle, a high level of detail may be required

Box 16.1 Ratio of Richest to Poorest, Selected Countries

	1987	2000
Brazil	26:1	29:1
Mexico	16:1	17:1
Guatemala	30:1	16:1
China	8:1	8:1
France	6:1	6:1
Jordan	6:1	6:1
Thailand	8:1	8:1
Canada	5:1	5:1
United States	9:1	9:1

Source: *United Nations Human Development Report 2000* (New York, 2003), Table 13. The table uses data on income and consumption to calculate an approximate representation of how much richer the wealthiest 20 per cent of the population are than the poorest 20 per cent. The higher the ratio, the more inequality there is in income distribution in that nation.

to ensure calculations are completed correctly. However, more often than not, the audience needs only to understand the approximate value, the order of magnitude, or a number in comparison to other numbers to gain a full understanding of your point. For example, is it more powerful to say that 'On 12 October 1999 the United Nations estimated that . . .

- the world surpassed 6 billion in population', or
- the world surpassed 6,132,544,000 in population'?

The second may be more accurate,[1] but the word 'billion' is more powerful and is more likely to be remembered by an audience than a string of numbers.

This balance between accuracy and meaning is of critical importance. As a presenter, ensure that the presentation focuses on the viewers or readers and their understanding of the numbers, and ensure that your numbers are correct. However, do not overemphasize the numbers themselves and your ability to present complex information. The point of the presentation is audience understanding, not numerical one-upmanship.

Necessary Numbers

A final issue with presenting numbers is whether the actual statistical data are necessary, or if a word statement would be a better choice. For example, if your objective is to sell peppermint gum, which statement would be more powerful?

- Exactly 53.52 per cent of survey respondents preferred peppermint gum over spearmint gum.
- The majority of survey respondents preferred peppermint gum over spearmint gum.

While presenting a specific number is accurate, words like 'majority' are easily understood and create a more powerful impression on the audience. Other words, like 'marginal difference', 'a fraction of', and 'an insignificant amount' create pictures for the audience that may be lost if the actual data values are used. Again, consider your purpose and the meaning you are trying to impart to the audience. One or two well-chosen words can ensure the audience fully understands the point of your presentation. At the same time, you always should strive for honesty and forthrightness in presenting statistical data. In the above example, after all, close to 47 per cent of respondents preferred spearmint gum, which, depending on the sample size, is a significant percentage. And if only 50.2 per cent preferred peppermint gum, you surely will be misleading an audience to speak of the 'majority' who prefer peppermint—most people, except at election time in two-party states, do not think of a 'majority' as this narrow.

Presenting Numbers for Understanding

Edward Tufte (2001)[2] coined the term 'chartjunk' to refer to unnecessary, useless, or obscuring elements in presenting statistical information or numbers in a document. If you intend to include numbers in a presentation or document, ask yourself one simple question: Does the inclusion of this information add to the audience's knowledge? If *yes*, ensure that the numbers are presented in a clear, easily understood fashion. If *no*, take the numbers out. A simplified approach is always preferred to an overly complex, incomprehensible stew of data.

It is important to note that 'simplified' should not be equated with 'dumbed down'. As was discussed in earlier chapters on writing for different audiences, an effective presentation is one that is understood, not one that is devoid of any intellectual meaning. Carefully consider if you are using numbers to assist the audience in understanding, if your motivations lean more towards obfuscation (to attempt to conceal key meanings) or pseudo-intellectualization (to prove how smart you are), or if you are simply presenting the data because you collected them as part of your research. The latter tends most often to be the case in university settings, where many an audience is brought to boredom by slide after slide of irrelevant data and the detailed discussion of statistical testing.

Norms for Numbers

When in doubt on presenting a document using numbers, consult with accepted reference guides (in many universities, APA style is the accepted norm, but confirm the style source with your institution or department).

In most style guides, the accepted practice is to express numbers less than 10 in words and numbers above as figures. Again, this can become confusing if more than a couple of numbers are contained in a sentence such as 'please reference Part 26, section 35, subsection two of the Act with focus on row three of Table 35.1 in the third column on page 159.' A general rule here is to use Arabic numerals for all numbers per se, even those below 10. Also, if possible, break apart the sentence to make the meaning clear.

Other generally accepted norms are:

- Never begin a sentence with numerals: '27 students attended the presentation.' Either spell out 'Twenty-seven' or restructure the sentence to begin with text: 'Among the attendees were 27 students.' Another frequently used option is to use a semicolon following what was the preceding sentence (unless, of course, your '27 students' began the paragraph).
- Use numbers and words when they better express a figure: $35 million is easier to understand than $35,000,000 for most readers.
- Do not use an apostrophe to pluralize a year—that is, it is the 1960s, not the 1960's.
- Do not list more than three numbers in a sentence (or on a slide). Clarity is lost when too many numbers are presented in a short span.

As a general rule, be logical and consistent in the use of numbers and statistics, and consult a style guide to be certain that the presentation of numbers meets established standards. A good on-line source is <www.askoxford.com>. Finally, as with most rules, logic should prevail over blind adherence to accepted practice. Focus on audience understanding and consistency and your presentation will be successful.

Chapter Review

This chapter examined issues surrounding the presentation of numbers. Again, the greatest issue is writing so the target audience will understand the information being presented. Consider the level of numeracy of the audience before presenting any numerical information.

Review Questions and Activities

1. What are the three rules for using numbers?

2. Seek out a good example and a less-than-stellar example of documents that present numbers. What factors add to the readability of the first document? What makes the second difficult to understand?

3. List five 'norms for numbers' and research other standards for the presentation of numbers in recognized style guides. Are there any surprises in the information?

4. When is a high level of detail required in presenting numbers? When does detail get in the way of presenting information?

5. Seek out spurious comparisons published in newspapers or magazines. What basis for comparison is used? Why?

Chapter Seventeen

Basics of Mapping

Maps are a familiar tool, used as a reference and a means of understanding the relationships among features on the Earth's surface. Ancient maps (possibly 40,000 years old) have been found in Africa, carved into rock faces by nomadic peoples and likely used to mark animal migrations. Maps on clay tablets carved by Babylonians (believed to show landownership) have been discovered, dating from 2300 BC. Maps of the world as it was understood have been found dating as far back as 600 BC. Although the origins of these ancient maps are not always fully known, the maps reveal important information about the societies of the map makers and their views of the world.

The evolution of settlements beginning more than 10,000 years ago brought gatherer societies to more settled agricultural societies, the relationships among cities and towns became important, as did trade routes and, in some societies, landownership.

Advances in astronomy and surveying, along with increased world exploration, added greater detail and accuracy to maps, and by the Middle Ages highly detailed (although strongly ethnocentric) maps of the world were being created to aid in navigation and determine territories. The invention of the printing press assisted mapping just as it benefited the production of text: a map could be replicated and made available to a much wider audience at a more reasonable cost.

Much of ancient mapping has been lost, as drawings carved into wood or traced on animal skins were not preserved over time. However, many examples of maps from the Middle Ages forward still exist, and the invention of the printing press allowed many copies to survive. A wealth of mapping from the last 200 years (pre-GIS) remains readily available and is much treasured by universities and collectors.

Technically, a map is a spatial depiction of data, with graphics depicting features or distributions on the Earth's surface. The relationship among these features or distributions can be analyzed by the map reader. A map might show landforms, average temperatures across an area, climate zones, or annual rainfall, or it might illustrate features related to the distribution of humans on the Earth's surface, such as density, voting patterns, religious identification, or other aspects of customs and norms.

But maps are more than that: the map is a 'telling' and an art. The map maker uses his or her skills to create a representation of elements on the Earth's surface, depicting what is important to the map maker. The map tells much about the map maker's background, knowledge level, biases, and beliefs.

At the same time, a map is not reality. It is important to understand that the map is simply the map maker's interpretation of the world as he or she knows it, produced with available technology and biased by the map maker's perceptions. The choice of data, the map maker's methods and methodology, and the techniques used all impact the map as it is seen by the map viewer.

Depending on the type of map produced, the map maker may calculate with indisputable accuracy the location of each map element. Or the map may be seen as art, with cartographic precision sacrificed for beauty. A third option is also available: the map maker may choose to manipulate the data presented to attach a prescribed meaning. That is, features may be emphasized or hidden, colour choices may be carefully made, and the chosen scale may be selected with awareness to expose or hide information on the map.

The Use of Maps

Why review mapping at all? From a graphic perspective, a map is one of the most telling means of getting information to an audience. Maps are used in many disciplines to show data dispersions, rates of change, and also the location of phenomena. The map is not only used by geographers and cartographers, but also by:

- sociologists showing dispersions of homeless populations;
- philosophers tracing the dispersion of ideas;
- criminologists mapping crime rates by neighbourhood;
- political scientists illustrating voting patterns;
- psychologists examining distances to needed services.

In short, maps are used by every researcher across the social sciences and humanities.

A map can show much more than geographic location; it also can place value on that location. Whether it shows incidence of disease or ratings of happiness, a map is the tool used to show how the dispersion of any phenomenon relates across an area on the surface of the Earth. Understanding the power of the map and using it as a means of presenting information is an important tool for researchers in any discipline.

While there are myriad maps worthy of consideration, three amazing examples of cartographic ability are described below, illustrating the range of information that can be presented on a map.[1]

Minard's Map of Napoleon's Journey

Charles Joseph Minard's thematic map of Napoleon's 1812–13 march to Moscow has been described by Edward Tufte as 'the best statistical graphic ever drawn'.[2] Created in 1861, the map uses a proportional line (with the line width proportional to the number of men) to track the Grande Armée as it moved across Europe. The line narrows as the Armée advanced, and branches off as troops left the main battalion. The same proportional line shows the retreat. The line width illustrates the colossal deterioration of the Armée, as terrible cold (the temperature is also marked on the base of the map), hunger, and the futility of the attempt decimated the troops.

The latitude and longitude of the quest are also shown on the map, as are key dates. An astounding amount of data are presented in a clear and easily understood form in this map.

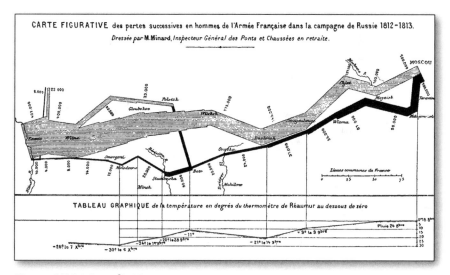

Figure 17.1 Minard's Map

Map of the World

While the project was ultimately unsuccessful, the vision of map makers to create an International Map of the World at a scale of 1:1,000,000 stands as one of the most optimistic in cartographic history. Beginning with discussions in the 1890s and world conferences and meetings of the International Geographical Congress in the early decades of the twentieth century, the project sought to create a small-scale topographic map of the world that would be uniform in scale, lettering, layers, and colours. It was expected that about 1,500 sheets would be required to span the earth's surface. More developed regions in the world agreed to provide the mapping for developing areas: Europe for Africa and the US for South America.

Two world wars slowed the production of the map, and by the 1960s key cartographers such as Arthur Robinson (Box 17.3) criticized the map as a waste of resources. Most nations lost interest, as advances in cartography led to the ability to create maps of greater importance to the map makers. The project foundered due to the lack of a co-ordinating body and any real need to complete the project.

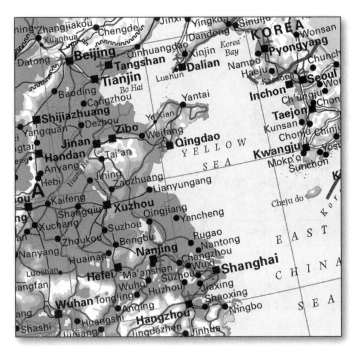

Figure 17.2 Map of the World: China
Source: Patrick Wiegand, *Oxford Student Atlas* (Oxford: Oxford University Press, 2006), copyright © Oxford University Press. Reprinted by permission of Oxford University Press.

Turner's Map of 'Life in LA'

In 1979, Eugene Turner, a geographer from California State University, Northridge, produced a map that used symbols to depict levels of affluence, unemployment rates, urban stresses, and population for an area within Los Angeles. The symbols used were 'Chernoff faces', a classification system developed by Herman Chernoff in 1973 to encode complex multivariate data on simple cartoon faces.

The map relies on innate human abilities to detect small variations in facial expressions in order to create an understanding of the relationships among facial features. The map clearly illustrates the concepts important to the map maker in a way that is both simple and complex. It stands as a brilliant map in its ability to present data in a way that provides meaning to the map viewer.

Maps are a graphic means of presenting an idea—a map can show not only landforms, but also the destruction of an army, the failed achievement of a lofty goal, or the feelings and emotions of residents in a way much more powerful than words.

Maps can be of great utility to researchers in the social sciences and humanities. In your next research project, consider if the addition of a map would provide greater clarity for readers or a new way of understanding the information you are presenting.

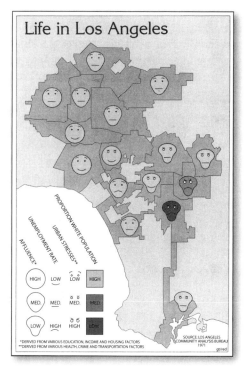

Figure 17.3 Turner's Map of Los Angeles
Source: Reprinted by permission from Eugene
Turner, PhD, California State University.

Types of Maps

Maps as a tool for communication have been critically important to the evolution of human civilization. A map is more than a way of presenting information; it is a way of recording features, routes, and places important to peoples and societies. On countless occasions, for example, Aboriginal people drew maps in sand or snow, or on birchbark or animal hide, to help early European explorers find their way in what today is Canada. A map allows the analysis of spatial patterns, comparing one place on the Earth's surface to another to find suitable lands for habitation and development. A map also provides information on societies by illustrating myriad details on people and places (such as changes in place names following conquests). Generally, maps serve three purposes:

1. *Wayfinding*. Certain maps are used for navigation: highway maps, train maps, and sailing charts are used to find the way from one place to another. The map may show more than one route, places of interest, and variations in travel paths (such as primary and secondary roads, or paved and gravelled road surfaces). The map viewer can begin at any point on the map and travel to any other, reuse the map for different trips, or use the map for different purposes (a tourist map can be used to find both well-known locales and to discover hidden places).

2. *Analysis*. Maps present data: the distance between places, the elevation of a mountain, and the slope of a hillside. A map is a shorthand reference for analyzing the Earth's surface. A map can show easier routes for hiking, give a comparative analysis of foreshore erosion between two points in time, and contain data that could otherwise only be obtained by direct field observation. The map is both a means of recording data and a way of allowing the map viewer to further his or her understanding the data.

3. *Understanding*. A map shows the spatial relationship of places on the Earth. While a place may be 'near' by a latitude/longitude calculation, a map can show the true association among places and destinations. Given that most maps represent the world at a smaller than actual scale (there are few 1:1 scale maps, as they would not be very useful for wayfinding, analysis, or understanding), a map allows us to see the world in a different, and more encompassing, way than we are capable of doing in person.

At a broad level, maps encountered by the professional are generally one of the following:

1. General purpose or *reference maps* are designed to be used to present information with technical accuracy for the map user's information

2. Special-purpose maps or *thematic maps* are designed to emphasize a single feature or 'theme'.

3. *Concept maps* are intended to illustrate an idea or model for an area, which can appear to be highly accurate. However, the intent of a concept map is to present imagery and fiction, not accuracy or the way things are at present.
4. *Aerial maps* are the product of photography, such as the satellite images produced by the National Aeronautics and Space Administration (NASA) in the US and the many images available from the National Snow and Ice Data Center (NSIDC) in Colorado.

Reference Maps

Reference maps are used to identify relationships among places: how roads connect in a city; the locations of bodies of water; the boundary of a municipality. The map is intended to be used as a technically accurate tool to allow the map reader to navigate across a portion of the Earth's surface. Whether it is a street map of Manhattan or a cartographically accurate depiction of the tributaries of the Amazon River, a reference map illustrates the spatial relationship of elements, ideally at a level or scale that is useful to the map reader.

Reference maps are interpreted by map readers as providing impartial, unbiased information on geographical features. The location of towns and villages, roads, topographic lines, and forms of vegetation are expected to be truthful and accurate: one expects that a published road map will accurately show the means of moving from A to B.

However, a map is merely a representation of reality, and is the product of the greater or lesser objectivity of the map maker and the limitations of selected data, as perceived by the map viewer. The process of simplifying features on the Earth,

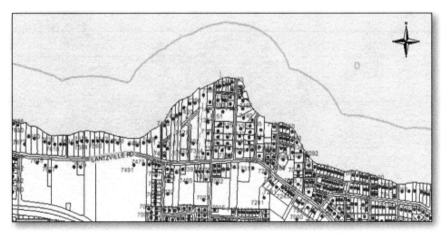

Figure 17.4 Reference Map: House Number Map of Lantzville, BC
Source: District of Lantzville.

determining an appropriate scale, selecting line widths and symbols, and placing them on the map will highlight some features and obscure others. Sometimes, data selection will inadvertently produce misinformation, as can occur with road maps when a road is planned (and mapped) but not constructed.

Even an aerial photo, what most would consider a fairly accurate depiction of reality, loses information in the translation of a three-dimensional world into a two-dimensional photo. Again, the reference map is but a representation of reality, and can never fully represent the world as it exists. Understanding this important caveat is critical for anyone relying on maps to move about the Earth's surface.

Thematic Maps

Thematic maps illustrate data or information related to a particular subject matter, and show the spatial distribution of a theme. Topics are limitless, from the distribution of population densities and plant species to snack-food consumption among university students. While thematic maps generally focus on a single issue, multivariate maps may also be created that illustrate the interaction among more than one variable. The thematic map is not intended to be used for wayfinding; instead, the map tells a story about a phenomenon, a means of understanding how an issue is distributed across the space of the mapped area.

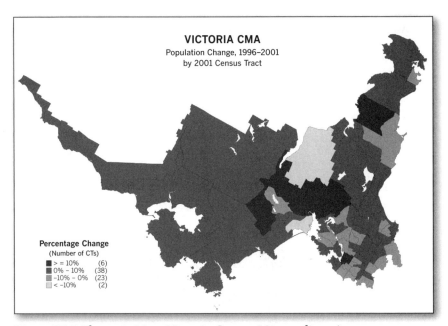

Figure 17.5 Thematic Map: Victoria Census Metropolitan Area
Source: Statistics Canada, at: <www.statcan.ca/bsolc/english/bsolc?catno=92F0173X2001935>.

Thematic maps do not necessarily provide accurate cartographic depictions of the world. In thematic mapping, the map maker may choose to soften or shift geographic features to better illustrate the point of the map, or use unusual colour schemes to emphasize the map maker's thesis.

A thematic map may use different techniques to represent data. Same-sized points may be used to show the density of a distribution, with each point representing the same number of occurrences (for example, one point may represent 10 people with a specific characteristic). Areas with a high number of occurrences would show a high density of points; areas with a low number would show only a few dispersed points.

The map could also use points that are proportional symbols, where points of different sizes are used to represent the number of occurrences. It is important to note that the size of the point may not be geometrically accurate—the map maker may manipulate the comparative size of points in a range to ensure that map viewers understand the mapped relationship. The map maker may use symbols that are not proportionally accurate but give the reader a *perception* of a relationship.

A thematic map may use colour or shading to represent data (referred to in some circumstances as *choropleth maps*, from the Greek 'choro', meaning area, and 'pleth', meaning value or data). Adherence to traditions in cartography, such as deeper colours for denser distributions of phenomena or the use of green for lands and blue for water, assist in the viewer's interpretation of the data. These maps normally illustrate a different colour for a range of occurrences, such as population in different age cohorts or the number of occurrences of a phenomenon. Thus, the thematic map is intended to provide a visual message to the viewer, one that hopefully is easily and accurately interpreted (that is, meeting the intentions of the map maker).

Concept Maps

Most often used in development proposals, a concept map shows an *intended* use of an area. They are a type of map often used by professionals in land development and architecture, and frequently encountered in urban life by residents in growing and changing environments.

A concept map is intended to do what its name implies: illustrate a proposal or concept that does not exist in reality. Although the map may appear highly detailed and contain elements that are an accurate representation of selected features on the landscape, the map represents only the map maker's depiction of a possible future.

Too often, concept maps are considered to be as geographically accurate as reference maps or to be as full of information as thematic maps. A concept map should not be given undue authority: it is a means of providing a visual of a proposed or intended future, and should be used for no other reason than understanding a proposal.

The peculiarity of concept maps is in the perception of the map viewer: because the map is an ancient and trusted source of transmitting information, viewers often equate a map with veracity. However, from the perspective of the map maker, the concept map is intended only to provide an idea or representation of a possible outcome for a development. In reality, the development may proceed in an entirely different manner, far removed from the concept presented.

Just as there are traditions in cartography, there are customs in concept mapping that are used to shape the viewer's interpretation of the map. The use of green shading denotes open space and parkland; water features, in blue, are an important element that is generally viewed favourably by the map viewer; and the depiction of trees and shrubs often leads to map viewers believing that represented greenery will either be retained on the site or planted. An educated map viewer understands that these are conventions that may or may not be realized in the final product.

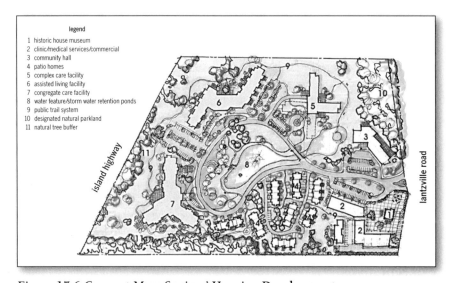

Figure 17.6 Concept Map: Seniors' Housing Development
Source: Reprinted with permission from James Dickinson, Lantzville, BC, and Chow Low Fleischauer Architects, Victoria, BC.

Aerial Maps

Other maps encountered by the professional include satellite and photo-generated mapping—Google Earth and other on-line mapping systems have revolutionized the look of maps. Water appears as it does in reality, and buildings and structures rise from the landscape. The map viewer can move down streets and turn corners, becoming a part of a kinetic map experience that is as much created by the map viewer as the map maker. Satellite photos and images of Canada's Far North, for example, have been instrumental in our understanding of the extent to which climate change has had an impact on the ever-shrinking ice cover in this region, which has serious social, environmental, economic, and political implications not only for the Inuit but for all Canadians.

It will be interesting to see if the use of highly interactive on-line mapping, GPS units, and hand-held computers will fully replace the paper map as a tool for navigating and understanding the world. Some researchers suggest that the two-dimensional paper map will become a historic artifact, prized by museums and collectors but of limited usefulness. Others would suggest that the map will persist as humans treasure the experience of holding and viewing a map. It is likely that the future of mapping is somewhere between these disparate views: the paper map will persist, but the use of hand-held devices will continue to advance and lead map viewers to previously unimagined views of the world.

Figure 17.7 Aerial Map: Manhattan
Source: Google Earth.

Features of the Map

A map is a process of transforming the physical reality found on the Earth's surface or the imagery in the mind of the map maker into a picture that speaks to the map viewer. The map maker selects the features that will be mapped, those that will be emphasized, the scale of the map, the classification system that will be used, and the elements that will be left off the map.

As the map is the creation of the map maker, there are myriad ways to present the same information on a map. While it may be difficult to determine the single 'right' way to illustrate a concept or data, missing or incorrect mapping elements will make a map more difficult to understand. Ultimately, the success of the map will be measured in its ability to convey the intended message to the map viewer. A 'good' map (meaning an effective map) is one that contains the following features: title, legend, symbols, accurate data, direction arrow, scale, projection, sources, and text labels.

Box 17.1 Good Titles/Bad Titles

Two of the following titles are effective and two are not—which are the better titles?

1. Population
2. 2006 Population by Census Tract, Maple Creek, Saskatchewan
3. Roads, Community Centres, Arenas, Parks, Trails, ParticiPaction tracks, Sewer Rights of Way and Water Easements in Medicine Hat, Alberta
4. Community Services in Flin Flon, Manitoba—2008

Titles 2 and 4 are more effective—they provide enough information for understanding but are not overly wordy.

Title

An effective title on a map is just as important as the title on a text or term paper. The map maker has one opportunity available to attract the map reader's attention and ensure the reader understands the purpose and usefulness of the map.

An effective title represents the content of the map. It is clear, descriptive, and short. The title should be located in a prominent location on the map—generally at the top or bottom edge of the map, with standardization of the title location if a series of maps is produced. The reader then knows what to expect of the map and can focus on the meaning of the map instead of trying to search out or interpret a weakly worded title.

Legend

Maps display three types of data:

1. *Points* are distinct places or occurrences that can be accurately placed on the map, such as cities, eagle trees, and incinerators.
2. *Lines* are features on the landscape with sharp and discernable boundaries, such as roads and river systems.
3. *Areas* are features that exist over the landscape, and may or may not have precise boundaries, such as the delineation of types of vegetation, a land-use zone, and the range of an animal species.

Box 17.2 Types of Mapped Data

Point: used to represent a discrete occurrence (with the point placed at the location of the data).

Line: used to represent linear features such as roads, rivers, and boundaries.

Area: used to represent data that occur over a dispersed area, such as the habitat of the Vancouver Island marmot.

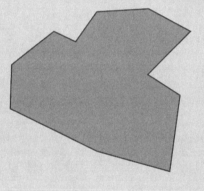

The legend on your map interprets the data for the map viewer. The range of colours used to represent data must be shown accurately in the legend, with the same colour value used on the map. Cartographic tradition would suggest that colour intensity increases with the density or frequency of the data (for example, higher population densities would be a darker or higher intensity colour than lower population densities). Data arrays are normally shown with the highest value at the top and the lowest at the bottom if arranged vertically, or lowest to the left and highest to the right if arranged horizontally.

If point data are used, the legend should be displayed in the same size as they appear on the map (see discussion on symbols, below).

Line widths and colours should be replicated exactly, as should the patterns and shading within areal polygons. The width of lines may vary to represent different features: the width of lines often varies on road maps, depending on road classifications, with the width of lines far exaggerated from reality (consider the width of a primary highway on a road map compared with the same road on an aerial photo).

The point of the legend is to assist viewers in interpreting the map—the viewers should not be forced to attempt to interpret the legend and then apply their 'best guess' to the features on the map. Give careful attention to the legend and ensure that all point, line, and area representations are accurately depicted.

Symbols

Symbols are generally used for illustrating point data or prominent features the map maker intends to highlight. A symbol is 'codespeak', a type of shorthand for conveying a message to the map viewer. As the map maker, you select the representations or codes that you will use on your map. For a map illustrating the location of potential incinerator locations in an area, which symbol would be most appropriate to mark each site:

- ● a point or dot;

 a flame; or a

 'thumbs-up' happy face?

By cartographic tradition, a point is most often used to represent a location, but the map maker has the latitude of choosing any symbol and can influence the map viewer's perceptions by selecting a symbol intended to invoke a particular reaction.

The use of carefully selected symbols also allows the map maker to draw attention to the highest-priority elements on the map. Again, as the map is simply a subjective representation of reality, not the reality it purports to represent, the map maker may use a symbol that either is or is not an accurate reflection of spatial reality—with intention or without. For example, urban settlements on a road map are often shown as proportional dots, with the size of dot increasing with population (and each dot representing a range of population sizes). The dot does not reflect reality; it is a cartographic convenience—most viewers would understand that urban settlements are not perfectly round, that a range of population sizes are represented by the same symbol, that the settlement would not exactly correspond to the location of the symbol, and even that settlements of equal population do not necessarily encompass the same area—some urban areas are sprawling, others are relatively compact. The size of the symbols does not represent actual conditions, as might be perceived in an aerial or satellite photo, but they provide useful information to the map viewer in an expected and easy-to-interpret format.

When you use symbols to denote locations on a map, if any type of symbol other than a technical or value-neutral form (such as a proportional dot) is used, be aware of the conventional meaning of the symbol. A black diamond has meaning for anyone who has been on a ski hill, but may not be known to non-skiers. Differences across cultures may also be a factor: a skeleton has different meaning in North America (associated with Halloween) and Mexico (when depicted in relation to a Día de los Muertos celebration). Give careful consideration to the use of symbols and be certain that the meaning is consistent with your intent as the map maker.

Accurate Data Representation

It is imperative that the map maker ensure the data used to create the map are sound and from reliable sources. Generally, information produced by government census agencies can be considered to be reasonably reliable; while there may be issues of absolute certainty with any number produced, these data are a best estimate for socio-demographic information. Data vetted through peer-reviewed academic publications can also be considered more reliable than numbers in a blog. If there is any doubt about the veracity of the data used, find other reliable sources that verify the information. Triangulating data (finding other sources that verify the data) is a good habit for any professional to adopt early in a career. Professional and personal embarrassment can result from using data that are, at best, inaccurate, or, at worst, absolutely false.

Maps themselves may be used as a reference by the map maker. The date of publication of the map is critically important: street patterns change; new lands are developed; earthquakes and tsunamis can significantly alter the Earth's surface; species habitats and the extent of old-growth forest are affected by changes in climate and by human activity. If the date of publication of a map is not known, or if it is possible that features on the map may have changed, be certain to research other mapping from reliable sources to verify the information on the map.

North Arrow

While seemingly a small part of a map, a north arrow is essential to orient the map viewer and critical if the map is to be used for wayfinding or navigation. In addition, a north arrow can be a logo for the map maker, a personal icon that depicts a map as his or her own. In large GIS departments, it is not uncommon for each map maker to create a personal north arrow, making it easy to trace a map back to its originator. Leaving a north arrow off a navigational map is a critical error in mapping, and certainly not one expected at a professional level.

Scale

A scale should be included on a map when it is accurate and the subject and form of the map support the use of the scale. Unless a map is at a 1:1 scale, which would mean that the map portrays a portion of the Earth of exactly the same size (that is, if you took a piece of paper and laid it on the Earth, then made a map that depicted exactly what was under that piece of paper), it is likely that your map is a *scaled* representation of reality.

A scaled map is a way of understanding the Earth. We cannot see great distances across the Earth's surface, hence reality on the map is reduced to a smaller scale so we can see the subject area in its entirety. However, with this reduction comes the issue of the selection of features to be shown or not included and the embellishment of features (road widths are shown much wider on highway maps than they actually are to assist in the viewer's ability to read the map).

The scale shows the relationship of distances on the map to actual distances on the Earth's surface. The map viewer can begin to understand the level of detail and cartographic accuracy by understanding the influence of selected scale on the map. Scale is commonly represented on a map in one of two ways.

Ratio scales illustrate a comparison. For example, a map with a scale of 1:1,000,000 is showing that one unit of measure on the map is equal to one million units of measure on the Earth's surface. If the measure used is centimetres, then one centimetre on the map equals one million centimetres (or one kilometre) on the Earth's surface. The value of a ratio is that any measure can be used. One cubit on the map is equal to one million cubits on the Earth's surface, and one inch is equal to one million inches.

The disadvantage of ratio scales is that the map cannot be altered or reproduced, and the scale is only accurate on the original map. If a ratio map is photocopied, the copier will slightly alter the size of the map and the relationship of features on the map (even if it is a 100 per cent copy). If the copy is enlarged or reduced, the scale is no longer accurate. Given the frequent need to photocopy mapping, the ratio scale is less useful than a second form of scale.

On a *line scale*, distances are marked along an extent and the map viewer can either estimate from this line or use a ruler to measure the distances between places. The map can be reduced or enlarged and the line scale remains true to the map.

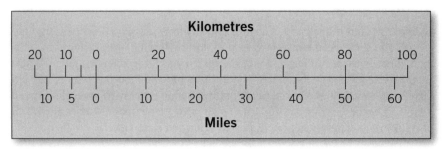

Figure 17.8 A Map Scale

One final note: a *large-scale map* is one that shows a small area of the world (so the features on the map are *large*), and a *small-scale map* depicts a much larger area (so the features on the map are *small*). Thus, on the same-size piece of paper, a large-scale map (say 1:5,000) would depict a small part of the Earth with great accuracy, while a small-scale map (1:1,000,000) would show a large part of the Earth with very little detail.

Cartographers do not agree on the precise point for classifying a map as large-scale, intermediate-scale, or small-scale. Generally, most would agree, however, that a map of 1:2,000 is a large-scale map, and 1:250,000 or smaller is a small-scale map.

Not all maps require a scale. A concept map that shows a redevelopment proposal for a city block may not have a scale attached if the buildings are at best hypothetical and intended only to offer an image of a potential outcome for the properties. A scale would also not be required on some thematic maps, where symbols are used to represent an issue on a familiar land form (North America) and the scale of the symbols is defined in the legend. For example, if proportional circles are used to represent numbers of viewers of professional hockey in major US markets, the underlying base map may not require a scale as the important information on the map is represented by the circles themselves. In either instance, the term 'not to scale' should be included at the base of the map or legend.

Projections

A map projection is required (and must be specified on the map) when all or a significant part of the spherical Earth is reduced to a two-dimensional, flat-surfaced map and the map is intended to be cartographically accurate.

Ideally, we would all carry globes with us at all times, at a sufficient scale to provide necessary information in the most geographically correct format. However, the size of globe needed to allow the viewer to find vehicle travel routes between Toronto and Montreal would be enormous, and unlikely to fit in an economy car. Given the impracticality of carrying a globe, map makers have long sought ways to accurately represent the surface of a sphere on a flat piece of paper.

If a map viewer were to peel off the surface of a globe and lay it out on a flat surface, cuts from the poles to the Equator would be required to enable this new depiction of the Earth. These cuts, or missing information, create an unappealing array, one that is not useful as a navigational source and one that does not assist in increasing the viewer's understanding of the relationships among features on the Earth's surface. Projections were developed by cartographers to fill in the blank spaces and make the world understandable.

Cartographers have created numerous ways to represent the Earth on a flat surface, each with its own issues of distortion. Some distort areas, others distort distances, and still others misrepresent the relationships among land masses. Knowing that different projections carry their own interpretive and representational advantages and disadvantages can assist the map viewer in interpreting data and understanding the relationships among places on a map. As a map viewer, it is important to check the source information on the map and confirm the projection used because this may have an impact on the map viewer's interpretation of the map.[3] The Robinson Projection (Box 17.3) is the projection map most viewers will recognize from texts in the social sciences and humanities.

Project or Source Information

What is the source of the base map and data? Documenting the source of the material is imperative to ensuring viewer confidence in the map. Too often, a map depicts information without accompanying text linking the map to a text or data reference. This information may be critical to the map viewer who wishes to further investigate the phenomena shown on the map. The source information should list the reference, date obtained (particularly if web-based), and credit the originators or authors.

Text Labels

Somewhere between labelling everything and labelling nothing is a happy place where meaningful text labels provide the map viewer with just enough information to understand and interpret the map. Use consistent fonts on your map, with

Box 17.3 Map Projections

The projection most students are familiar with is the *Robinson Projection*, named for Arthur H. Robinson (1915–2004), an American cartographer. He wrote more than 60 texts and articles, and coined the term 'map percipient' with co-author Barbara Petchenik, which considers how the viewer sees and understands the map.

The Robinson Projection was created in the early 1960s, with the intention of limiting the distortion found in other map projections. Through an extensive trial-and-error process, Robinson created a 'pseudo-cylindrical' projection, which he labelled an 'orthophanic' projection, but the name Robinson Projection is much more commonly used. The projection is unique for two reasons: the means of creating the map (though trial and error) is unusual, as is the emphasis on creating a visually appealing map.

no more than two different fonts (one for the title and one for labels). The use of more than two fonts can be distracting and look unprofessional. Font sizes should be consistent as well, with the same font size used for similarly weighted data.

The placement of text labels must also be by design, not default. Use caution in placing text labels to ensure that key features on the map are not obscured by the labels. As a map viewer, be aware that labels may be placed to hide features that the map maker wants to distract attention from; putting an urban place name across an undesirable feature (such as a landfill site) is fairly common in mapping. Be intentional in label placements to ensure that the prominent features of your map remain clearly visible and easily distinguished by the map viewer.

The Appearance of the Map

A map provides useful information, but it is also a picture that can achieve the status of an art form. What makes a map valuable? Often, the age of the map, the techniques used, and/or the importance of the map maker in the history of cartography or as a member of a significant expedition determines the value of a map, but sometimes the map simply is graphically beautiful and aesthetically pleasing. While creating beauty may not be the map maker's primary focus, an effective layout for a map adds to the map viewer's ability to interpret the information on the map, which is the ultimate goal of the map maker. Several criteria can add to the appearance of a map.

Highlighting. Numerous studies on eye movement suggest that humans do not view maps (or pages of information) in a logical or linear start-to-finish manner; instead, the eyes first move to those things with visual weight like titles, bright colours, or areas of high contrast. The viewer may do additional sweeps of the map, but this first overview is how the map viewer forms an impression of the content and usefulness of the map. As the map maker, it is imperative to ensure that those elements most important to understanding the map are highlighted, with correct meaning attached to this highlighting or emphasis.

Differentiation. An effective map clearly differentiates between the explanatory elements on the map. If proportional symbols are used, the difference in sizes must be easily determined by the map viewer. Differences in line widths should be obvious, as should the range of colours used to represent these elements. If it is likely that a map will be photocopied, test the colour range and symbols used to ensure that the copy adequately represents the differentiations on the map. The colours produced by different colour printers can also differ markedly, and grey tones can fade together, making map interpretation difficult.

Focus on the user. Perhaps most important is the tailoring of the map to the intended audience. If a map is intended to be a reference map to provide accurate information for the travelling public, then the map maker must be certain that the map is current and accurate—dated or incorrect information could create an inconvenience or real problems for the user. If the map is a thematic map on a specific issue, the map must accurately tell the story of that feature or distribution.

Clarity on the purpose of the map will greatly assist the map maker in designing the map. Before beginning any development, carefully consider the intended audience and their use of the product, then work to develop the most effective and useful map possible (within the context of available resources, software limitations, costs, and time constraints).

Data Collection and Accuracy

In most cases, the data you will use to produce a map will be from one of two sources: information you collect yourself from first-hand field experience (original source data), or information drawn from another source and used in your mapping (secondary source). For both sources, the use of data must consider:

- the means by which the data were collected; and
- the reason the data were collected.

For example, you may have reason to complete an on-site visual survey of a 10-hectare lot. In your field notebook, you carefully note the location of the watercourse and major trees, and even map out an area of what appears to be Garry oak habitat. You are collecting the information to add to the knowledge of land uses on the property in preparation for a concept map. Back in the office, you are asked to turn this information into a GPS-referenced map that will be used to determine surveyed boundaries. Obviously, the on-site visual survey for a concept map, however thoroughly completed, cannot be used to determine legal lot lines. How you collected the data, and why, does not suit the secondary use of the data.

The same caution should be used when considering the use of secondary source information. Know how the information was collected, and for what purpose. Be certain you are not replicating errors in the data that were produced as part of the original information collection.

In mathematics, an error is the difference between a calculated quantity and its true value. In mapping, detecting an error is more complicated. For example, unless a map is at a 1:1 scale, all information is simplified and generalized. A river is shown in an approximate location, a road width does not match reality. Are these errors? The map is not correct in this sense; however, the 'error' is a product of reducing the complexity of reality to aid the map viewer's understanding. While the accuracy of a map can be verified through a review of the GPS data and the criteria used for producing the map, it is important that the map viewer understands that the map itself can never be without 'error' from a purely mathematical definition of the word.

Most mapping errors are not created from a lack of data, but are simply mistakes in data entry or interpretation. Unless the map maker is intentionally placing false information on the map (for whatever reason, such as advancing a particular political agenda), these errors can be detected and corrected.

Practical tip. In a geological sense, most landforms do not change quickly over time. Mountains remain in place, the ocean stays in roughly the same location, and rivers follow their course. However, cataclysmic events can change the shape of the world. The profound changes to areas around the Indian Ocean following the earthquake and tsunami in December 2004 illustrate how the Earth can change over time. The 'human footprint' on the Earth's surface can also change rapidly, as is witnessed in the growth of urban areas across the planet. If the features being mapped are subject to change, and the data being used are dated, it would make sense to seek out the most current secondary source, or, better yet, complete new primary data collection.

What Is the Purpose of Your Map?

The ability to technically produce a map that contains all required elements (title, scale, legend, source, north arrow, and accurate text labels, as previously discussed) is an important skill for any professional. Just as important is knowing *how* to use the map to present your ideas and objectives in a way that will present *your* reality of the map.

As noted by Mark Monmonier, the foremost expert in understanding the map as a tool of communication, 'A single map is but one of an indefinitely large number of maps that might be produced for the same situation or from the same data. . . . A good map tells a multitude of little white lies; it suppresses truth to help the user see what needs to be seen.'[4] In other words, the map is not reality, but it is your reality as the map maker. The elements selected for the map (the features that are chosen or not chosen) are a way of transforming the physical reality of the Earth onto the map.

Of the four types of maps previously discussed (reference, thematic, aerial, and concept maps), the latter most strongly manipulates reality. The power of a concept map is that it uses a technique for presenting visual information that people normally see as true—we tend to believe the information we see on a map, and this gives a concept map a much higher level of credence than would be given to a text describing the concept. The difference is that the text is 'just words' while the concept map presents a reality on the Earth's surface. Even more believable are the concept maps that show an imagined reality as though it already exists through the placement of photo-generated buildings and structures on an aerial photo of a development site. New techniques are being evolved daily to allow the viewer to fly over, around, and through a development, to experience the development from eye level at a human scale. Increasing the 'believability' of the concept map is a powerful tool in creating a new truth for the viewer.

A map is a summary of facts—those facts seen as most important by the map maker. The purpose of the map remains a communication with the map viewer, through a relationship that trusts in the map viewer's ability to understand and interpret the map. As with any text-based document, the purpose of the map is to tell a story or create an impression. However, the paucity of explanatory text on the map requires greater clarity from the map maker and greater awareness from the map viewer on the data sources, methods used, and the map maker's intentions for the map.

As noted by Robinson and Petchenik, the relationship between the map maker and the map viewer is worthy of further study: 'the nature of the map as an image and the manner in which it functions as a communication device between the cartographer and percipient need much deeper consideration and analysis than they have yet received.'[5] Along with this analysis should be an increased understanding of the value of the map itself. A map is a tool of communication that is as important to the evolution of human society as the written word. Maps have led us to new places, and allowed us to return to old haunts. Thematic maps allow us an understanding of relationships that cannot be seen on the Earth's surface: relationships in language, culture, religion, and meaning. Maps give us locations, show change over time,

Box 17.4 Five Principles of Cartographic Design

Concept before Compilation

Without a grasp of the concept and audience, the whole of the design process is negated. Once the concept and audience are understood, no design or content feature will be included which does not fit. Design the whole before the parts, the parts encompass the whole.

The cartographic design process has two stages: 1) concept and parameters, and 2) detail in execution. When developing the concepts and parameters, design, devise, design again. Audience is important, design for the user first, the user last. What does the map user want from this map? What can the map user get from this map? Is that what they want? If the map were a building, it should have a strong enough foundation that it should not fall over. Just like a building's foundation, a map's foundation should be an integral part of the map but not the dominant feature of the map.

Hierarchy with Harmony

Important features must look important, and the most important features should look the most important. If you removed the title from the map, could the map user quickly determine what the theme (most important thing) of the map was? Lesser map features have their place and should serve to complement the design and the important features. Every feature on a map from the whole to the part, and all the parts, should be contributing to the whole. Harmony is to do with the whole map being happy with itself. Successful harmony leads to repose. Perfect harmony leads to neutral bloom. Harmony is subliminal.

Simplicity from Sacrifice

Great design tends towards simplicity (Jacques Bertin). Content may determine scale or scale may determine content, and each determines the level of generalization or sacrifice. It is not what you put in it that makes a great map, but what you take out. The map design stage is complete when you can take nothing else out. This allows the map to focus the attention of the map user like running the film of an explosion backwards. All possibilities rush to one point. They become the right point. This is the objective with thematic cartography. This is the cartographer's skill.

Maximum Information at Minimum Cost

How much information can be gained from a map at a glance? How long will the map user realistically look at a map? Functionality is more important than utility. Good design makes utility functional. All designs are a compromise, just as a new born baby is a compromise between its father and mother. The park which makes a map special often only comes when the map is complete.

Engage the Emotion to Engage the Understanding

Design with emotion to engage the emotion. Only by feeling what the map user feels can we see what the map user sees. Good cartographers use cartographic fictions, impressions, and illusions to make a map. All of these have emotive contents. The image is the message. Good design is a result of the tension between the environment (the facts) and the cartographer. Only when the map user engages the emotion, the desire, will they be receptive to the map's message.

Design uses aesthetics but the principles of aesthetics are not those of design. Cartographers are not just prettying maps up. The philosophy is simple, beauty (aesthetics) focuses attention. Focusing the attention of the map user is the purpose of cartographic design!

Source: Society of Cartographers, University of Glasgow, 1999, at: <cartotalk.com/index.php?showtopic=311>.

and quantify spatial distributions. No other method is available to give a visual representation of the processes that operate to change the planet; without maps, the ability to contextualize information would be sadly reduced.

The Role of the Map Maker

Ultimately, the map is the project of the map maker. As the maker, you decide on the content of the map, font used, colours, layout, data sources, the classification system on the map, and the scale or projection. The most important consideration is being aware of the purpose of the map, and working to achieve this purpose to the maximum level of your cartographic abilities.

As noted previously, a map is a representation of elements on the Earth's surface, an illustration of a theme, or an imagined concept. The map tells much about the map maker's background, knowledge level, biases, and beliefs. Be certain that the map produced is an accurate representation of you, the map maker, and your intentions for the project.

Chapter Review

This chapter reviewed the map as a presentation tool and its usefulness to researchers across the social sciences and humanities. While the map is most closely associated with research in geography, its usefulness to other researchers also was discussed. Four types of maps were considered—reference maps, thematic maps, concept maps, and aerial maps, e.g., satellite photography and imaging. The various important features of maps were examined, as were the aesthetics of map making and the relationship between the map maker and the intended users of the map.

Review Questions and Activities

1. What are the key elements of a good map? In any textbook with mapped information, review the maps to see if all the elements are present in each map. What is missing?

2. What is the purpose of a map? How could a map be used in your next research project or paper?

3. Why is it important to understand the means of data collection on your map?

4. Finish this statement:
   A large-scale map _____ *while a small-scale map* _____.

5. Considering the three famous maps presented in this chapter, research other famous mapping. What criteria define your selection of important or interesting maps?

Afterword

By the end of this book, you will have gained new skills in writing, research, public presentations, and the use of illustrations and mapping illustrations. As a new professional, you will have new ideas to present to the world, and improved communication skills are the fastest way to ensure that your ideas are recognized and implemented.

Strive to build new skills at every opportunity—avail yourself of every text you can access, attend professional development courses, observe the skills of seasoned professionals, and acquire new skills to supplement your writing, performance skills, and use of visuals, from presentation props, PowerPoint, and flip charts to map making. The best writers are those who read a great deal; the best performers are those who study the performances of others; and the best map makers have spent many hours poring over the maps that others have made.

Notes

Chapter 1

1. Now in its fourth edition, Strunk and White's *Elements of Style* continues to influence the way information is portrayed by students and others interested in improving their ability to communicate. E.B. White (the author of *Charlotte's Web* and a student of Strunk's at Cornell University) was added as a co-author in 1957 after White added a new chapter to 'the little book'.
2. There are other texts that students may also find useful, such as Sheridan Baker's *Complete Stylist*, *Practical Stylist*, or *Complete Stylist and Handbook*, and many universities publish their own style guides.
3. George Orwell, *Politics and the English Language* (London: Horizon, 1946).
4. European Union, Directorate-General for Translation, *Fight the Fog: How to Write Clearly*, 2. This document is available through the European Union's website, at: <ec.europa.eu/translation>.
5. Ibid., 9.
6. That is, members of the Canadian Institute of Planners and particularly the Planning Institute of British Columbia would understand that British Columbia's Local Government Act sets out requirements for Official Community Plans, Development Permits, Development Variance Permits, and the Board of Variance.

Chapter 2

1. When time is an issue, a fast method of brainstorming is to use the Internet. Enter key terms in your search engine and see what comes up. The sites may or may not be relevant, but they may help to expand your thinking. Of course, a full-scale literature review is a good method of understanding previous research in your field of study—literature reviews are considered in this chapter.
2. Many organizations use APA format, a style of citation developed by the American Psychological Association. The use of citations is discussed later in this textbook, and citation manuals are available (either as textbooks or online) for the writer. Other formats can be seen in professional journals. The most important point is to select a format and be consistent.
3. George Orwell, *Politics and the English Language* (London: Horizon, 1946).
4. Joseph A. Ecclesine, 'Advice to scientists—in words of one syllable', *American Journal of Economics and Sociology* 24, 3 (1965): 271. For a useful discussion of this source, see Richard Nordquist, 'Richard's Grammar and Composition Blog', at: <grammar.about.com/b/2007/11/30/keep-it-brief.htm>.

Chapter 3

1. K.G. Jones and M.J. Doucet, 'The big box, the flagship, and beyond: Impacts and trends in the Greater Toronto Area', *Canadian Geographer* 45, 4 (2001): 404–512.
2. An Ethics Review Committee is either a university-wide or department-specific committee that reviews all research proposals conducted by students and staff to ensure that the work meets established standards. Most such committees seek to verify that the research will 'do no harm' to the subjects of research, whether human, animal, or even inanimate. The committee will look for specific information on the questions that will be asked, the sampling methods that will be used, how results will be recorded, and the proposed end use of the study information. In general, no research involving human or animal subjects can be conducted as part of university-based research without the expressed written permission of an Ethics Review Committee.

Chapter 4

1. Briefing notes are discussed here, but you also may be asked to provide a briefing paper (with a longer, more comprehensive format) or an oral briefing (where you verbally update the decision-maker).
2. 'Cost-effective' does not always mean that you have the lowest bid. Most organizations retain the right not to accept the lowest or any bid if they feel the requirements of the project cannot be met.

Chapter 5

1. A. Grafton, *The Footnote: A Curious History* (Cambridge, Mass.: Harvard University Press, 1997), 8.
2. Miguel Roig, 'When college students' attempts at paraphrasing become instances of potential plagiarism', *Psychological Reports* 84, 3 (1999): 973–82.

Chapter 6

1. As noted by James Crawford in 'Endangered Native American languages: What is to be done, and why?', *Bilingual Research Journal* 19, 1 (1995): 17–38, many of these lost languages will be those spoken by indigenous populations. As speakers age and youth are decreasingly likely to speak the language, words and meaning are lost. Eventually, no new speakers are added, the language becomes moribund, and eventually vanishes, taking with it cultural expressions and meanings that could only be transmitted in that language.

Chapter 7

1. Philosopher David Hume in the 1700s coined the term 'inductive fallacy', which means it is illogical to assume that any given pattern of events will follow the same pattern in the future. The researcher must be aware that what *should* follow is not necessarily what *will* follow, particularly if human subjects are part of the research equation.
2. Robert Merton, *Social Theory and Social Structure* (New York: Free Press, 1968).
3. See, e.g., Rae Bridgman, Sally Cole, and Heather Howard-Bobiwash, eds, *Feminist Fields: Ethnographic Insights* (Peterborough, Ont.: Broadview Press, 1999).
4. See, e.g., Juanne Nancarrow Clarke, *Health, Illness, and Medicine in Canada*, 4th edn (Toronto: Oxford University Press, 2004), 309.
5. Durkheim supported the use of empirical methods (the scientific method, as discussed in this chapter) to study social phenomena. He removed the focus from the individual and instead sought out the structures that determined the behaviours of the individual (see Kenneth Thompson, *Emile Durkheim* [London: Tavistock, 1982], 11–18). Durkheim saw these structures as dynamic, but the structures could be understood through applying rigorous methods, enabling the researcher to find the connections among layers in the social structure.
6. T.S. Palys, *Research Decisions: Quantitative and Qualitative Perspectives* (Scarborough, Ont.: Thomson Nelson, 2003), 10.
7. See, e.g., ibid. For a good discussion of qualitative and quantitative methods in the social sciences, see Bruce Arai, 'Research Methods', in Lorne Tepperman and Patrizia Albanese, *Sociology: A Canadian Perspective*, 2nd edn (Toronto: Oxford University Press, 2008), ch. 2. A useful text is Earl Babbie, *The Basics of Social Research* (Toronto: Wadsworth Thomson Learning, 2002). For a historical view of methodology specific to geography, and discussion that favours a qualitative approach, see William Norton, *Human Geography*, 6th edn (Toronto: Oxford University Press, 2007), ch. 2, especially 76–8 and, for useful additional references, 82–3.

Chapter 8

1. As web mapping develops, the quality and level of detail is rapidly improving. Street-level web cameras have been introduced in major cities to provide real-time views of streets and locations. This may be a valuable resource for your observational research.
2. To view the Hanover College Psychology Department's website, 'Psychology Research on the Net', visit <psych.hanover.edu/research/exponent.html>.
3. Even the most experienced field researchers, on occasion, will find themselves in the field without a tape measure. As a backup, practise striding in one-metre increments and use yourself as a measuring stick—if you are exactly two metres tall, you can be used to record with reasonable accuracy the size of a tree or building. Know the length of your arm span or the measure from the tip of your middle finger to your elbow. Know the dimensions of your vehicle— how many metres in length and height? Be able to accurately estimate the size

of an acre or hectare. While accurate measuring devices are preferred, use your observational skills to understand a site if no other option is available.

Chapter 9

1. Ayelet Meron Ruscio, Timothy A. Brown, Wai Tat Chiu, Jitender Sareen, Murray B. Stein, and Ronald C. Kessler. 'Social fears and social phobia in the United States: Results from the National Comorbidity Survey Replication', *Psychological Medicine* 38, 1 (2008): 15–28.
2. The reaction to these latter stimuli is often referred to as a 'startle reaction'. A falling infant (or adult) will tense the muscles, often pulling in the extremities to protect the body. The eyes will open wide, then squint, and heart rate and blood pressure will increase dramatically. Whether or not the person is actually under threat, the reaction is the same.

Chapter 10

1. M. McKeown and M. Curtis, *The Nature of Vocabulary Acquisition* (Hillsdale, NJ: Lawrence Erlbaum, 1987).

Chapter 11

1. With credit to Sherrin Western, a professional speaker and principal of Shervin Communications Inc. in Vancouver.
2. The study of non-verbal communication is not new; Charles Darwin published *The Expression of the Emotions in Man and Animals* in 1872. However, much of the work on 'body language' began in the mid-twentieth century. In one influential study, anthropologist Ray Birdwhistell estimated that we can recognize about 250,000 human facial expressions. Many excellent texts have been written on understanding non-verbal communication. See especially Nalini Ambady and Robert Rosenthal, 'Thin slices of expressive behaviour as predictors of interpersonal consequences: A meta-analysis', *Psychological Bulletin* 111, 2 (1992): 256–74; Robert A. Hinde, *Non-verbal Communication* (London: Cambridge University Press, 1972); Andrew Beck, Peter Bennett, and Peter Wall, *Communication Studies: The Essential Introduction* (New York: Routledge, 2005).

Chapter 14

1. Lynne Cooke, 'Eye tracking: How it works and how it relates to usability', *Technical Communication* 52, 4 (2005): 456–63, at: <www.ingentaconnect.com/content/stc/tc;jsessionid=48ti6i0ur77sd.alice>.
2. Current research into eye tracking suggests that patterns may be different for reading web pages. It appears that readers are more likely to read a web page in an 'F' pattern, scanning the headline or banner first, then skimming along the left-hand side of the page, reading the 'headlines' or highlighted text before viewing text that is lower in the text hierarchy.

Chapter 15

1. For further information, see the works of psychologist Albert Mehrabian, listed in the Bibliography, on different rates of retaining text vs visual presentations. Mehrabian found that retention rates could be 50–80 per cent higher through the use of a combination of presentation methods (text and visual) than through reading text alone.

Chapter 16

1. The UN has noted that, given inaccurate and non-existent census data in parts of the world, it is difficult to know precisely when the world passed the 6 billion mark; 12 October 1999 is used as a best guesstimate and as a means of making a point about population growth.
2. Tufte's *The Visual Display of Quantitative Information* (2001) should be on the bookshelf of every student and professional.

Chapter 17

1. Readers are asked to seek out their own examples of mapping achievement. Recommended texts for maps from the 1400s forward are David Bannister and Carl Moreland, *Antique Maps: A Collector's Guide*, 3rd edn (Oxford: Phaidon, 1993); David Buisseret, *The Mapmakers' Quest: Depicting New Worlds in Renaissance Europe* (Oxford: Oxford University Press, 2003).
2. Edward Tufte, *The Visual Display of Quantitative Information*, 2nd edn (Cheshire, Conn.: Graphics Press, 2001), 42.
3. For example, a Mercator projection causes greater distortion towards the poles; land masses such as Greenland are depicted as much larger than their actual size on the Earth's surface. For a good reference on projections, see the discussion in Quentin Stanford, *Canadian Oxford School Atlas*, 7th edn (Toronto: Oxford University Press, 1998).
4. Mark Monmonier, *How to Lie with Maps* (Chicago: University of Chicago Press, 1996), 23. This text should be on the desk of every student and professional in the social sciences.
5. Arthur H. Robinson and Barbara B. Petchenik, *The Nature of Maps* (Chicago: University of Chicago Press, 1976), 20.

Bibliography

Ambady, Nalini, and Robert Rosenthal. 1992. 'Thin slices of expressive behaviour as predictors of interpersonal consequences: A meta-analysis', *Psychological Bulletin* 111, 2: 256–74.

American Psychological Association. 2007. *APA Style Guide to Electronic References*. Washington: American Psychological Association.

Arai, Bruce. 2008. 'Research Methods', in Lorne Tepperman and Patrizia Albanese, *Sociology*, 2nd edn. Toronto: Oxford University Press, ch. 2.

Babbie, Earl. 2002. *The Basics of Social Research*. Toronto: Wadsworth Thomson Learning.

Baker, Sheridan. 1977. *The Practical Stylist*, 4th edn. New York: Harper & Row

Bannister, David, and Carl Moreland. 1993. *Antique Maps: A Collector's Guide*, 3rd edn. Oxford: Phaidon.

Beattie, Geoffrey. 2003. *Visible Thought: The New Psychology of Body Language*. London and New York: Routledge.

Beck, Andrew, Peter Bennett, and Peter Wall. 2005. *Communication Studies: The Essential Introduction*. New York: Routledge.

Behrens, Laurence. 2007. *A Sequence for Academic Writing*. New York: Pearson Longman.

Blanck, Peter David. 1993. *Interpersonal Expectations: Theory, Research, and Applications*. Cambridge: Cambridge University Press.

Bradbury, Andrew. 2006. *Successful Presentation Skills*. London: Kogan Page.

Bridgman, Rae, Sally Cole, and Heather Howard-Bobiwash, eds. 1999. *Feminist Fields: Ethnographic Insights*. Peterborough, Ont.: Broadview Press.

Buchanan, Elizabeth A. 2004. *Readings in Virtual Research: Ethics Issues and Controversies*. Hershey, Penn.: Information Science.

Buisseret, David. 2003. *The Mapmakers' Quest: Depicting New Worlds in Renaissance Europe*. Oxford: Oxford University Press.

Cesario, Joseph, and Tory Higgins. 2008. 'Making message recipients "feel right": How nonverbal cues can increase persuasion', *Psychological Science* 19, 5: 415–20.

Clarke, Cheryl. 2007. *Grant Proposal Makeover: Transform Your Request from No to Yes*. San Francisco: Jossey-Bass.

Clarke, Juanne Nancarrow. 2004. *Health, Illness, and Medicine in Canada*, 4th edn. Toronto: Oxford University Press.

Coates, Linda Jane. 2006. *Qualitative and Quantitative Research Methods Reader: A Canadian Orientation*. Toronto: Pearson Prentice-Hall.

Coley, Soraya M. 2000. *Proposal Writing*. Thousand Oaks, Calif.: Sage.

Cooke, Lynne. 2005. 'Eye tracking: How it works and how it relates to usability', *Technical Communication* 52, 4: 456–63, at: <www.ingentaconnect.com/content/stc/tc;jsessionid=48ti6i0ur77sd.alice>.

Craswell, Gail. 2005. *Writing for Academic Success: A Postgraduate Guide*. Thousand Oaks, Calif.: Sage.

Crawford, James. 1995. 'Endangered Native American languages: What is to be done, and why?', *Bilingual Research Journal* 19, 1: 17–38.

Creswell, John W. 2003. *Research Design: Qualitative, Quantitative, and Mixed Method Approaches*. Thousand Oaks, Calif.: Sage.

Delanty, Gerard. 2001. *Challenging Knowledge: The University in the Knowledge Society*. Buckingham, UK and Philadelphia: Open University Press and Society for Research into Higher Education.

DeMarrais, Kathleen Bennett. 2004. *Foundations for Research Methods of Inquiry in Education and the Social Sciences*. Mahwah, NJ: Lawrence Erlbaum Associates.

Dent, Borden D. 1996. *Cartography: Thematic Map Design*, 4th edn. Dubuque, Iowa: William C. Brown.

De Sausmarez, Maurice. 2001. *Basic Design: The Dynamics of Visual Form*. London: A. & C. Black.

Dreier, James Lawrence. 2008. *Contemporary Debates in Moral Theory*. Malden, Mass.: Blackwell.

Ebel, Hans Friedrich, C. Bliefert, and W.E. Russey. 2004. *The Art of Scientific Writing: From Student Reports to Professional Publications in Chemistry and Related Fields*, 2nd edn. Weinheim, Germany: Wiley-VCH.

Ecclesine, Joseph A. 1965. 'Advice to scientists—in words of one syllable', *American Journal of Economics and Sociology*, cited in Richard Nordquist, 'Richard's Grammar and Composition Blog', at: <grammar.about.com/b/2007/11/30/keep-it-brief.htm>.

Elderfield, John. 1999. *Body Language*. New York: Museum of Modern Art; dist. by H.N. Abrams.

Ely, Margot. 1997. *On Writing Qualitative Research: Living by Words*. London: Falmer Press.

European Union, Directorate General for Translation. *Fight the Fog: How to Write Clearly*. At: <ec.europa.eu/translation>.

Feldman, R.S., and Bernard Rime. 1991. *Fundamentals of Nonverbal Behavior*. New York: Cambridge University Press.

Fonstad, M., W. Pugatch, and B. Voght. 2003. 'Kansas is flatter than a pancake', *Annals of Improbable Research* 9, 3: 16–18.

Frank, Francine Harriet Wattman. 1989. *Language, Gender, and Professional Writing: Theoretical Approaches and Guidelines for Nonsexist Usage*. New York: Modern Language Association, Commission on the Status of Women in the Profession.

Fuller, Steve. 2006. *The Philosophy of Science and Technology Studies*. New York: Routledge.

Gibaldi, Joseph. 1998. *MLA Style Manual and Guide to Scholarly Publishing*. New York: Modern Language Association.

Giltrow, Jane. 2005. *Academic Writing: An Introduction*. Peterborough, Ont.: Broadview Press.

Grafarend, Erik W. 2006. *Map Projections: Cartographic Information Systems*. Berlin: Springer.

Grafton, A. 1997. *The Footnote: A Curious History*. Cambridge, Mass.: Harvard University Press.

Hall, Judith A. 1984. *Nonverbal Sex Differences: Communication Accuracy and Expressive Style*. Baltimore: Johns Hopkins University Press.

Harley, J.B. 1989. 'Deconstructing the map', *Cartographica* 26, 2: 1–20.

Harner, James L. 2000. *On Compiling an Annotated Bibliography*. New York: Modern Language Association.

Hinde, Robert A. 1972. *Non-verbal Communication*. London: Cambridge University Press.

Holdstein, Deborah H. 2001. *Personal Effects: The Social Character of Scholarly Writing*. Logan: Utah State University Press.

Holland, Jeremy. 2005. *Methods in Development Research: Combining Qualitative and Quantitative Approaches*. Rugby, Warwickshire, UK: ITDG.

Holloway, Brian R. 2003. *Proposal Writing across the Disciplines*. Upper Saddle River, NJ: Prentice-Hall.

Horgan, Paul. 1973. *Approaches to Writing*. New York: Farrar, Straus and Giroux.

Jones, K.G., and M.J. Doucet. 2001. 'The big box, the flagship, and beyond: Impacts and trends in the Greater Toronto Area', *Canadian Geographer* 45, 4: 404–512.

Kalbfleisch, P.J. 1993. *Interpersonal Communication: Evolving Interpersonal Relationships*. Mahwah, NJ: Lawrence Erlbaum Associates.

———— and M.J. Cody. 1995. *Gender, Power, and Communication in Human Relationships*. Mahwah, NJ: Lawrence Erlbaum Associates.

Kantola, Steve. 2006. *How to Write and Deliver Great Speeches: The Toastmasters International Guide to Public Speaking*. Mill Valley, Calif.: Kantola Productions.

Karpf, Anne. 2006. *The Human Voice: How This Extraordinary Instrument Reveals Essential Clues about Who We Are*. London: Bloomsbury.

Keohane, Nannerl O. 2006. *Higher Ground: Ethics and Leadership in the Modern University*. Durham, NC: Duke University Press.

Knight, Robert M. 1998. *The Craft of Clarity: A Journalistic Approach to Good Writing*. Ames: Iowa State University Press.

Krygier, John. 2005. *Making Maps: A Visual Guide to Map Design for GIS*. New York: Guilford Press.

Lam, Nina. 1983. 'Spatial interpolation methods', *American Cartographer* 10, 2: 129–49.

LeBlanc, Linda. 1996. *Writing a Proposal : A Step-by-Step Guide*. Edmonton: Literacy Services of Canada.

Lee-Treweek, Geraldine, and Stephanie Linkogle, eds. 2001. *Danger in the Field: Risk and Ethics in Social Research*. London: Routledge.

Limb, Melanie. 2001. *Qualitative Methodologies for Geographers: Issues and Debates*. London: Arnold.

Lipson, Charles. 2006. *Cite Right: A Quick Guide to Citation Styles—MLA, APA, Chicago, the Sciences, Professions, and More*. Chicago: University of Chicago Press.

McKeown, M., and M. Curtis. 1987. *The Nature of Vocabulary Acquisition*. Hillsdale, NJ: Lawrence Erlbaum Associates.

McLuhan, Marshall. 1964. *Understanding Media: The Extensions of Man*. New York: McGraw-Hill.

Madonik, Barbara G. 2001. *I Hear What You Say, but What Are You Telling Me? The Strategic Use of Nonverbal Communication in Mediation*. San Francisco: Jossey-Bass.

Mallon, Thomas. 1989. *Stolen Words: Forays into the Origins and Ravages of Plagiarism*. New York: Ticknor & Fields.

Mandel, Steve. 1987. *Effective Presentation Skills*. Los Altos, Calif.: Crisp Publications.

Marczyk, Geoffrey R. 2005. *Essentials of Research Design and Methodology*. Hoboken, NJ: John Wiley & Sons.

Marino, Kim. 1989. *The College Student's Resume Guide: Writing Your Own Professional Resume*. Santa Barbara, Calif.: Tangerine Press.

Maxwell, Joseph Alex. 2005. *Qualitative Research Design: An Interactive Approach*. Thousand Oaks, Calif.: Sage.

Mehrabian, Albert. 1972. *Non-verbal Communication*. Piscataway, NJ: Aldine Transaction.

——— and Alan D. Baddeley. 1979. *The Psychology of Memory*. New York: Perseus Books Group.

Mertens, Donna M. 2005. *Research and Evaluation in Education and Psychology: Integrating Diversity with Quantitative, Qualitative, and Mixed Methods*. Thousand Oaks, Calif.: Sage.

Merton, Robert. 1968. *Social Theory and Social Structure*. New York: Free Press.

Miller, Delbert Charles. 2002. *Handbook of Research Design and Social Measurement*. Thousand Oaks, Calif.: Sage.

Monmonier, Mark S. 1982. *Computer-Assisted Cartography: Principles and Prospects*. Englewood Cliffs, NJ: Prentice-Hall.

———. 1996. *How to Lie with Maps*. Chicago: University of Chicago Press.

——— and George A. Schnell. 1988. *Map Appreciation*. Englewood Cliffs, NJ: Prentice-Hall.

Nagin, Carl. 2003. *Because Writing Matters: Improving Student Writing in Our Schools*. San Francisco: Jossey-Bass.

Nardi, Peter M. 2006. *Doing Survey Research: A Guide to Quantitative Methods*. Boston: Pearson/Allyn & Bacon.

Noble, David F. 2005. *Gallery of Best Resumes for People without a Four-Year Degree: A Collection of Quality Resumes by Professional Resume Writers*. Indianapolis: JIST.

Nola, Robert. 2007. *Theories of Scientific Method: An Introduction*. Montreal and Kingston: McGill-Queen's University Press.

Norton, William. 2007. *Human Geography*, 6th edn. Toronto: Oxford University Press.

Orwell, George. 1946. *Politics and the English Language*. London: Horizon.

Palys, T.S. 2003. *Research Decisions: Quantitative and Qualitative Perspectives*. Scarborough, Ont.: Thomson Nelson.

Perrin, Kathleen Ouimet. 2008. *Ethics and Conflict*. Sudbury, Mass.: Jones and Bartlett.

Pfeiffer, William S. 2000. *Proposal Writing: The Art of Friendly and Winning Persuasion*. Upper Saddle River, NJ: Prentice-Hall.

Philippot, Pierre, R.S. Feldman, and Erik Coats. 1999. *The Social Context of Nonverbal Behavior*. Cambridge: Cambridge University.

Resnick, Elizabeth. 1987. *Graphic Design: A Problem-solving Approach to Visual Communication*. Englewood Cliffs, NJ: Prentice-Hall,

Rhind, D., ed. 1999. *Geographic Information Systems: Principles and Technical Issues*, vol. 1, 2nd edn. New York: John Wiley and Sons.

Robinson, Arthur H., Joel L. Morrison, Phillip C. Muehrcke, A. Jon Kimerling, and Stephen C. Guptill. 1995. *Elements of Cartography*, 6th edn. New York: John Wiley & Sons.

———— and Barbara B. Petchenik. 1976. *The Nature of Maps*. Chicago: University of Chicago Press.

Roig, Miguel. 1999. 'When college students' attempts at paraphrasing become instances of potential plagiarism', *Psychological Reports* 84, 3: 973–82.

Rubin, Herbert J. 2005. *Qualitative Interviewing: The Art of Hearing Data*. Thousand Oaks, Calif.: Sage.

Ruscio, Ayelet Meron, Timothy A. Brown, Wai Tat Chiu, Jitender Sareen, Murray B. Stein, and Ronald C. Kessler. 2008. 'Social fears and social phobia in the United States: Results from the National Comorbidity Survey Replication', *Psychological Medicine* 38, 1: 15–28.

Russell, Nick. 2006. *Morals and the Media: Ethics in Canadian Journalism*. Vancouver: University of British Columbia Press.

Sennett, Richard. 2006. *The Culture of the New Capitalism*. New Haven: Yale University Press.

Shank, Gary D. 2006. *Qualitative Research: A Personal Skills Approach*. Upper Saddle River, NJ: Pearson Merrill Prentice-Hall.

Shinew, Dawn M. 2003. *Information Literacy Instruction for Educators: Professional Knowledge for an Information Age*. Binghamton, NY: Haworth Information Press.

Silverman, David. 2007. *A Very Short, Fairly Interesting and Reasonably Cheap Book about Qualitative Research*. London: Sage.

Smith, Charles Kay. 1974. *Styles and Structures: Alternative Approaches to College Writing*. New York: Norton.

Somerville, Margaret. 2006. *The Ethical Imagination: Journeys of the Human Spirit*. Toronto: Anansi.

Speck, Bruce W. 1988. *Grading Student Writing: An Annotated Bibliography*. Westport, Conn.: Greenwood Press.

Stanford, Quentin. 1998. *Canadian Oxford School Atlas*, 7th edn. Toronto: Oxford University Press.

Strunk, William, and E.B. White. 1979. *The Elements of Style*. New York: Macmillan.

Swan, Michael, and Catherine Walter. 2001. *The Good Grammar Book*. Oxford: Oxford University Press.

Taschner, Rudolf J. 2007. *Numbers at Work: A Cultural Perspective*. Wellesley, Mass.: A.K. Peters.

Thompson, Kenneth. 1982. *Emile Durkheim*. London: Tavistock.

Tufte, Edward R. 1990. *Envisioning Information*. Cheshire, Conn.: Graphics Press.

———. 1997. *Visual Explanations: Images and Quantities, Evidence and Narrative*. Cheshire, Conn.: Graphics Press.

———. 2001. *The Visual Display of Quantitative Information*, 2nd edn. Cheshire, Conn.: Graphics Press. First published 1983.

Turabian, Kate L. 2007. *A Manual for Writers of Research Papers, Theses, and Dissertations: Chicago Style for Students and Researchers*. Chicago: University of Chicago Press.

Valiela, Ivan. 2001. *Doing Science: Design, Analysis, and Communication of Scientific Research*. Oxford: Oxford University Press.

Williams, Robin. 1994. *The Non-Designer's Design Book: Design and Typographic Principles for the Visual Novice*. Berkeley, Calif.: Peachpit Press.

Witz, Marion. 1997. *Stand Up and Talk to 100 People (and Enjoy It!)*. Toronto: McLeod Publishing.

Wood, Julia T. 1999. *Gendered Lives: Communication, Gender, and Culture*. Belmont, Calif.: Wadsworth.

Woodmansee, Martha. 1994. *The Construction of Authorship: Textual Appropriation in Law and Literature*. Durham, NC: Duke University Press.

Woodward, David. 1992. 'Representations of the World', in Ronald F. Abler, Melvin G. Marcus, and Judy M. Olson, eds, *Geography's Inner Worlds: Pervasive Themes in Contemporary American Geography*. New Brunswick, NJ: Rutgers University Press, 50–73.

Zimbardo, Philip G. 2008. *The Lucifer Effect: Understanding How Good People Turn Evil*. New York: Random House.

Index